Thermoplasmonics

Guohua Liu

# Thermoplasmonics

## From Principles, Materials and Characterization to Engineering Applications

Springer

Guohua Liu
Key Laboratory of Power Station Energy
Transfer Conversion and System, Ministry
of Education
North China Electric Power University
Beijing, China

ISBN 978-981-97-8331-1 ISBN 978-981-97-8332-8 (eBook)
https://doi.org/10.1007/978-981-97-8332-8



This Springer imprint is published by the registered company Springer Nature Singapore Pte Ltd.
The registered company address is: 152 Beach Road, #21-01/04 Gateway East, Singapore 189721, Singapore

# Preface

Thermoplasmonics is a highly promising and rapidly expanding subfield of plasmonics. Under irradiation, plasmonic nanostructures exhibit strong light absorption, transforming into nanoscale heat sources that can be precisely controlled. By utilizing light to heat nanostructures, researchers have unlocked a wide range of applications and possibilities for enhanced light–matter interactions at sub-wavelength scales.

This book is built on the recent advancements in understanding thermoplasmonics and highlights the exciting new directions that are shaping this field. I will delve into the fundamental principles of plasmonic theory, thermodynamics, and thermal-induced processes to lay a solid foundation for the following exploration. Material selection and rational structural design are crucial aspects in maximizing the efficiency of energy conversion in thermoplasmonics. I discuss the various materials used in thermoplasmonics and delve into common synthesis methods, providing insights into achieving optimal thermoplasmonic properties. Characterization technologies play a vital role in understanding and modeling thermoplasmonic systems. I will explore analytical and computational approaches as well as nanoscale thermometry techniques used to characterize these systems accurately. In addition, this book also showcases the wide applications of thermoplasmonics in different fields, including life, optoelectronic information science, energy, and chemistry, environmental and material sciences. Each application will be accompanied by an in-depth discussion of the unique role of thermoplasmonics within that particular field and the associated benefits. Finally, I additionally analyze research trends and provide insights into using thermoplasmonics to achieve improved performance. Challenges faced by researchers in this field and future perspectives will be highlighted, ensuring that readers gain a comprehensive understanding of both the current state and potential future directions of thermoplasmonics research.

This book serves as an essential resource for scientists, engineers, and researchers interested in exploring the fascinating world of thermoplasmonics. It is my hope that this comprehensive framework will inspire further exploration and innovation in this rapidly evolving field. I extend my gratitude to all the researchers and scientists who have contributed to the development of thermoplasmonics and around the world who have given me the encouragement and insight needed to put the book together.

Beijing, China
August 2024

Guohua Liu

# Contents


**1 Introduction** .......... 1
References .......... 3

**2 Fundamental Principles of Thermoplasmonics** .......... 7
2.1 Plasmonic Theory .......... 7
2.1.1 Metal Electromagnetism .......... 7
2.1.2 Plasmon Resonance Modes .......... 10
2.1.3 Quality Factors .......... 14
2.2 Thermodynamics .......... 17
2.2.1 Heat Generation .......... 17
2.2.2 Heat Transfer .......... 19
2.2.3 Temperature Profile .......... 21
2.2.4 Heating Scheme .......... 22
2.3 Thermal-Induced Physics .......... 24
2.3.1 Thermooptical Effects .......... 24
2.3.2 Phase Change and Stress Wave .......... 26
2.3.3 Mass Transport .......... 28
2.3.4 Chemical Reaction .......... 30
References .......... 32

**3 Thermoplasmonic Materials** .......... 41
3.1 Material Selection .......... 41
3.1.1 Metal Nanostructures .......... 43
3.1.2 Metallic Derivatives/Semiconductor Crystals .......... 44
3.1.3 2D Materials .......... 45
3.2 Material Design .......... 46
3.2.1 Size Effect .......... 47
3.2.2 Shape Effect .......... 49
3.2.3 Intraparticle Coupling .......... 51
3.2.4 Interparticle Coupling .......... 53
3.2.5 Active Coupling .......... 54
3.3 Material Fabrication .......... 57

3.3.1 Top-Down: Lithography ... 57
3.3.2 Bottom-Up: Colloidal Synthesis ... 60
3.3.3 Self-assembly and Hybrid Approaches ... 63
References ... 66

**4 Characterization Techniques** ... 77
4.1 Analytical Model ... 77
4.1.1 Electrostatic Limit ... 78
4.1.2 Retardation Corrections ... 79
4.1.3 Plasmon Frequency ... 80
4.2 Computational Approaches ... 82
4.2.1 FDTD Method ... 82
4.2.2 Discrete Dipole Approximation ... 85
4.2.3 Boundary Element Methods ... 86
4.2.4 Multiple Multipole Methods ... 87
4.3 Nanoscale Thermometry ... 87
4.3.1 Thermal Radiation Spectroscopy ... 88
4.3.2 Raman Spectroscopy ... 89
4.3.3 Fluorescent Molecular Probes ... 91
4.3.4 Nanoprobes ... 93
4.3.5 Surrounding Variation ... 97
References ... 101

**5 Engineering Applications** ... 107
5.1 Life Science ... 107
5.1.1 Photothermal Therapy ... 107
5.1.2 Drug and Molecular Delivery ... 110
5.1.3 Cell Biology ... 112
5.1.4 Polymerase Chain Reaction ... 114
5.1.5 Fight Against COVID-19 ... 115
5.2 Optoelectronic Information Science ... 117
5.2.1 Thermoplasmonic Sensor ... 117
5.2.2 Photothermal Imaging ... 119
5.2.3 Photoacoustic Imaging ... 120
5.2.4 Heat-Assisted Magnetic Recording ... 122
5.2.5 Thermoplasmonic Infrared Detector ... 123
5.3 Energy Science ... 125
5.3.1 Thermophotovoltaics ... 125
5.3.2 Thermoplasmonic Desalination ... 127
5.3.3 Solar Thermal Collector ... 129
5.3.4 Radiative Cooling/Energy-Saving Window and Anti-icing/Fogging ... 131
5.3.5 Thermoplasmonic Motor ... 133
5.4 Chemical Science ... 134
5.4.1 Chemical Synthesis ... 134
5.4.2 Solar Hydrogen ... 136

5.4.3 Solar $CO_2$ Reduction ... 138
5.4.4 Nitrogen Fixation ... 140
5.5 Environmental Science ... 142
5.5.1 Water Purification ... 142
5.5.2 Sterilization and Disinfection ... 144
5.5.3 Removal of Nano/Microplastics ... 145
5.6 Material Science ... 147
5.6.1 Plasmonic Colloidal Synthesis ... 147
5.6.2 Thermoplasmonic Welding ... 149
5.6.3 Additive Manufacturing ... 151
References ... 153

**6 Summary and Outlook** ... 179

# Chapter 1
# Introduction

Thermoplasmonics is an increasingly important technology in scientific research due to its ability to control temperature at the nanoscale [1–4]. It operates based on the unique physics of plasmonic nanostructures interacting with light. Plasmonic materials, often in the form of metallic nanoparticles, manifest localized surface plasmon resonances (LSPRs) arising from the collective electron oscillations in response to incoming electromagnetic waves [5]. Upon illumination, plasmonic nanostructures absorb photons which trigger collective oscillations of the electrons with the frequency of incident light. This synchronized electron movement significantly intensifies the electric field near the particle surface, creating localized electromagnetic field intensification referred to as hot spot [6, 7]. This intense hot spot generated by plasmonic nanoparticles serves as an efficient source of heat. The uptake of light energy by plasmonic particles initiates nonradiative decays, including electron–electron scattering and electron–phonon coupling, leading to the conversion of photon energy into thermal energy [8, 9].

The localized heat generated in thermoplasmonics is of particular importance. The rise in temperature can be controlled by fine-tuning variables like the choice of plasmonic materials, particle dimensions and morphology, properties of the incident light, surrounding conditions, and the application of external stimuli [10]. Different materials exhibit unique plasmonic properties, including resonance wavelength, scattering, and absorption efficiency [11, 12]. By selecting the appropriate material, it is possible to tune the plasmonic response [13–17]. Changing the sizes or dimensions of nanostructures alters their LSPR characteristics, affecting the absorption and scattering of light [18, 19]. It is possible to precisely control the amount of absorbed energy and the resulting heat generation by adjusting the incident light properties, such as wavelength, intensity, and duration. Heat dissipation from plasmonic nanoparticles is also influenced by surrounding medium. By modifying the thermal conductivity and heat transfer characteristics, the heat distribution and dissipation efficiency can be manipulated [20–22]. Furthermore, external stimuli, such as electric or magnetic fields, also can be utilized to actively control thermoplasmonics.



G. Liu, *Thermoplasmonics*, https://doi.org/10.1007/978-981-97-8332-8_1

The application of these fields can induce changes in the plasmonic nanostructures' properties, leading to altered heat generation and distribution profiles. This aspect opens up possibilities for dynamic control of thermoplasmonic effects [23–25].

In the nanoplasmonic systems, heating quantification is crucial in examining the experimental observations. However, quantifying the sample heating is very difficult [12, 26]. Nanothermometry is an emerging tool in measuring the thermal effect. While the complexity of nanoscale heating poses challenges to be addressed [27, 28]. It is important to note that both experimental and simulation methods should complement each other in understanding the thermoplasmonics [6, 29, 30]. Furthermore, the optics community is more conversant with propagation, interference, and diffraction phenomena and often has less familiarity with the diffusion laws. The thermoplasmonic effects gained considerable attention within plasmonics and nanooptics, but ambiguous debates persist on how to accurately describe these phenomena [31, 32]. At present, the community is still not fully understood and consensual regarding the impact of nanoscale photothermal processes [1]. Therefore, understanding thermoplasmonics requires interdisciplinary collaboration and a more comprehensive understanding of both plasmonics and thermal physics.

The distinct characteristics of plasmonic materials, combined with their ability to intensify the local electromagnetic fields, are crucial in leveraging thermal energy for a diverse range of applications [2–4, 33, 34]. The generated heat can induce various effects depending on the specific application. For example, metal nanoparticles exhibiting resonances in the near-infrared spectrum can be engineered to bind selectively to cancer cells. When irradiated with resonant light, these particles become heated, effectively eradicating tumor cells while sparing the surrounding healthy tissue [35, 36]. Similarly, the selective heating of plasmonic particles is explored for its potential in laser-assisted release of conjugated substances, paving the way for targeted drug delivery [37, 38]. In energy conversion, the heat produced can bolster solar cell efficiency by aiding in the charge separation and transport. Such progress holds the promise to boost the effectiveness of solar energy systems, contributing to sustainable energy strategies [3, 39]. In sensing applications, the temperature change caused by plasmonic effects can be measured to detect and quantify analytes with high sensitivity [40, 41]. Furthermore, the controlled heat generation enables the fabrication of nanoscale materials and devices with enhanced performance [42–44]. Recently, the range of applications of thermoplasmonics has expanded beyond this initial suite of direct uses.

In this book, we overview the recent advancements in understanding of thermoplasmonics and highlight the exciting new directions that are shaping this field (Scheme 1). First, we delve into the essential principles, starting with concepts from plasmonic theory and thermodynamics and extending to thermo-induced processes. Substantial effort is then devoted to exploring various thermoplasmonic materials alongside prevalent fabrication techniques. We highlight strategies that guide the optimal selection of materials and rational design for enhanced energy conversion. Moreover, inventive methodologies for fabrication novel nanostructures are detailed, aiming at fine-tuning thermoplasmonic properties. We also discuss the advanced techniques used to characterize thermoplasmonics, including simple analytical and

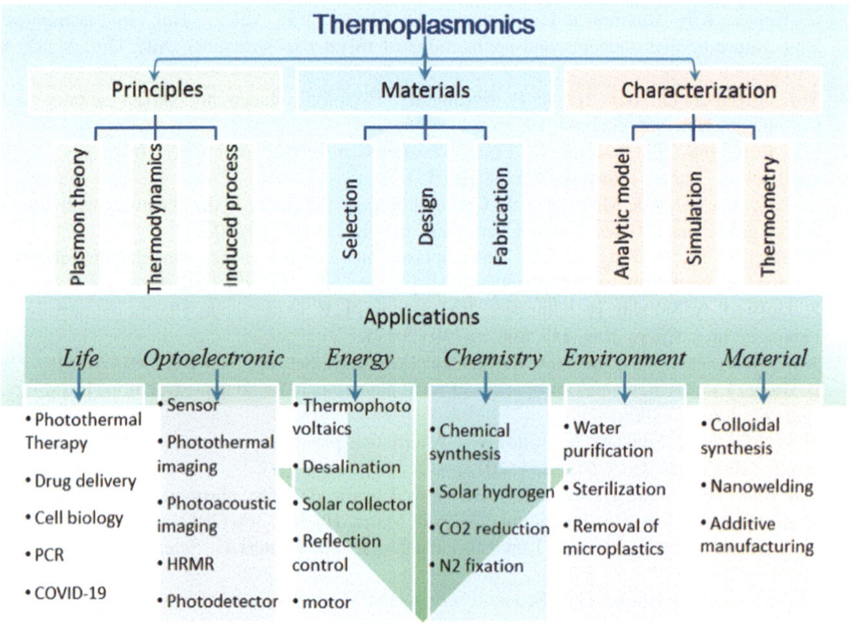


**Scheme 1** Structure of this book

computational models as well as nanoscale thermal measurement tools. Lastly, we succinctly depict the potential application of thermoplasmonics in areas critical to life, energy, environment, and materials. For each application, the unique role of thermoplasmonics on specific applications and their benefits wrought by thermoplasmonic innovation are elaborated. Finally, we delineate the prevailing challenges facing the field of thermoplasmonics and suggest prospective avenues for future development.

## References


1. G. Baffou, F. Cichos, R. Quidant, Applications and challenges of thermoplasmonics. Nat. Mater. **19**, 946–958 (2020)
2. V.T. Ruhoff, M.R. Arastoo, G. Moreno-Pescador, P.M. Bendix, Biological applications of thermoplasmonics. Nano Lett. **24**(3), 777–789 (2024)
3. G. Liu, J. Xu, T. Chen, K. Wang, Progress in thermoplasmonics for solar energy applications. Phys. Rep. **981**, 1–50 (2022)
4. S. Santoro, A.H. Avci, A. Politano, E. Curcio, The advent of thermoplasmonic membrane distillation. Chem. Soc. Rev. **51**, 6087–6125 (2022)
5. S.A. Maier, *Plasmonics Fundamentals and Applications*. (Springer Science+Business Media LLC, 2007)
6. G. Baffou, R. Quidant, Thermo-plasmonics: using metallic nanostructures as nano-sources of heat. Laser Photonics Rev. **7**, 171–187 (2012)

7. C. Kuppe, K.R. Rusimova, L. Ohnoutek, D. Slavov, V.K. Valev, "Hot" in Plasmonics: temperature-related concepts and applications of metal nanostructures. Adv. Opt. Mater. **8** (2019)
8. M.L. Brongersma, N.J. Halas, P. Nordlander, Plasmon-induced hot carrier science and technology. Nat. Nanotechnol. **10**, 25–34 (2015)
9. S.V. Boriskina, H. Ghasemi, G. Chen, Plasmonic materials for energy: from physics to applications. Mater. Today **16**, 375–386 (2013)
10. A. Rao, A.S. Iglesias, M. Grzelczak, Choreographing oscillatory hydrodynamics with DNA-coated gold nanoparticles. J. Am. Chem. Soc. **146**, 18236–18240 (2024)
11. B. Yang, C. Li, Z. Wang, Q. Dai, Thermoplasmonics in solar energy conversion: materials, nanostructured designs, and applications. Adv. Mater. **34**, e2107351 (2022)
12. L. Jauffred, A. Samadi, H. Klingberg, P.M. Bendix, L.B. Oddershede, Plasmonic heating of nanostructures. Chem. Rev. **119**, 8087–8130 (2019)
13. A.B. Urcan Guler, V.M. Shalaev, Refractory plasmonics. Science **344**, 263–264 (2014)
14. F. Wang, Y.R. Shen, General properties of local plasmons in metal nanostructures. Phys. Rev. Lett. **97**, 206806 (2006)
15. G.V. Naik, V.M. Shalaev, A. Boltasseva, Alternative plasmonic materials: beyond gold and silver. Adv. Mater. **25**, 3264–3294 (2013)
16. J.M. Luther, P.K. Jain, T. Ewers, A.P. Alivisatos, Localized surface plasmon resonances arising from free carriers in doped quantum dots. Nat. Mater. **10**, 361–366 (2011)
17. A. Boltasseva, H.A. Atwater, Low-Loss Plasmonic Metamaterials. Science **331**, 290–291 (2011)
18. M.B. Ross, M.G. Blaber, G.C. Schatz, Using nanoscale and mesoscale anisotropy to engineer the optical response of three-dimensional plasmonic metamaterials. Nat. Commun. **5**, 4090 (2014)
19. D.R. Ward, F. Huser, F. Pauly, J.C. Cuevas, D. Natelson, Optical rectification and field enhancement in a plasmonic nanogap. Nat. Nanotechnol. **5**, 732–736 (2010)
20. L.O. Herrmann, V.K. Valev, C. Tserkezis, J.S. Barnard, S. Kasera, O.A. Scherman, J. Aizpurua, J.J. Baumberg, Threading plasmonic nanoparticle strings with light. Nat. Commun. **5**, 4568 (2014)
21. L. Meng, R. Yu, M. Qiu, F.J. Garcia de Abajo, Plasmonic nano-oven by concatenation of multishell photothermal enhancement. ACS Nano **11**, 7915–7924 (2017)
22. D. Yoo, F. de León-Pérez, M. Pelton, I.H. Lee, D.A. Mohr, M.B. Raschke, J.D. Caldwell, L. Martín-Moreno, S.H. Oh, Ultrastrong plasmon-phonon coupling via epsilon-near-zero nanocavities. Nat. Photonics **15**, 125–130 (2020)
23. L.D. Sio, *Active Plasmonic Nanomaterials* (CRC Press, Taylor & Francis Group, LLC, 2016). International Standard Book Number-13: 978-981-4613-4601-4619
24. N. Jiang, X. Zhuo, J. Wang, Active Plasmonics: principles, structures, and applications. Chem. Rev. **118**, 3054–3099 (2018)
25. T. Ding, V.K. Valev, A.R. Salmon, C.J. Forman, S.K. Smoukov, O.A. Scherman, D. Frenkel, J.J. Baumberg, Light-induced actuating nanotransducers. Proc. Natl. Acad. Sci. U.S.A. **113**, 5503–5507 (2016)
26. G. Baffou, *Thermoplasmonics Heating Metal Nanoparticles Using Light* (Cambridge University Press, 2018). ISBN 978-971-108-41832-41834
27. M. Quintanilla, L.M. Liz-Marzán, Guiding rules for selecting a nanothermometer. Nano Today **19**, 126–145 (2018)
28. M. Da Browski, Y. Dai, H. Petek, Ultrafast photoemission electron microscopy: imaging Plasmons in space and time. Chem. Rev. **120**, 6247–6287 (2020)
29. G.B. Matthew Pelton, *Introduction to Metal-Nanoparticle Plasmonics* (Wiley, 2013), p. 275
30. G. Baffou, R. Quidant, F.J. Garcia de Abajo, Nanoscale control of optical heating in complex plasmonic systems. ACS Nano **4**, 709–716 (2010)
31. R. Chen, Z. Ren, Y. Liang, G. Zhang, T. Dittrich, R. Liu, Y. Liu, Y. Zhao, S. Pang, H. An, C. Ni, P. Zhou, K. Han, F. Fan, C. Li, Spatiotemporal imaging of charge transfer in photocatalyst particles. Nature **610**, 296–301 (2022)

32. O. Nicoletti, F. de la Pena, R.K. Leary, D.J. Holland, C. Ducati, P.A. Midgley, Three-dimensional imaging of localized surface plasmon resonances of metal nanoparticles. Nature **502**, 80–84 (2013)
33. X. Zhang, Y.L. Chen, R.S. Liu, D.P. Tsai, Plasmonic photocatalysis, reports on progress in physics. Physical Society **76**, 046401 (2013)
34. U. Das, R. Biswas, N. Mazumder, Elucidating thermal effects in plasmonic metal nanostructures: a tutorial review. Eur. Phys. J. Plus **137** (2022)
35. B. Luo, W. Wang, Y. Zhao, Y. Zhao, Hot-Electron dynamics mediated medical diagnosis and therapy. Chem. Rev. **123**, 10808–10833 (2023)
36. S.Y. Lee, M.J. Shieh, Platinum(II) drug-loaded gold nanoshells for chemo-photothermal therapy in colorectal cancer. ACS Appl. Mater. Interfaces. **12**, 4254–4264 (2020)
37. M.J. Mitchell, M.M. Billingsley, R.M. Haley, M.E. Wechsler, N.A. Peppas, R. Langer, Engineering precision nanoparticles for drug delivery. Nat. Rev. Drug Discovery **20**, 101–124 (2021)
38. E. Blanco, H. Shen, M. Ferrari, Principles of nanoparticle design for overcoming biological barriers to drug delivery. Nat. Biotechnol. **33**, 941–951 (2015)
39. J.G. Smith, J.A. Faucheaux, P.K. Jain, Plasmon resonances for solar energy harvesting: a mechanistic outlook. Nano Today **10**, 67–80 (2015)
40. G. Qiu, Z. Gai, Y. Tao, J. Schmitt, G.A. Kullak-Ublick, J. Wang, Dual-functional Plasmonic photothermal biosensors for highly accurate severe acute respiratory syndrome coronavirus 2 detection. ACS Nano **14**, 5268–5277 (2020)
41. E. Haque, M. Anwar Hossain, Y. Namihira, F. Ahmed, Microchannel-based plasmonic refractive index sensor for low refractive index detection. Appl. Opt. **58**, 1547–1554 (2019)
42. J. Lee, I. Lee, T.S. Kim, J.Y. Lee, Efficient welding of silver nanowire networks without post-processing. Small **9**, 2887–2894 (2013)
43. S. Han, S. Hong, J. Ham, J. Yeo, J. Lee, B. Kang, P. Lee, J. Kwon, S.S. Lee, M.Y. Yang, S.H. Ko, Fast plasmonic laser nanowelding for a Cu-nanowire percolation network for flexible transparent conductors and stretchable electronics. Adv. Mater. **26**, 5808–5814 (2014)
44. L. Wang, Y. Feng, Z. Li, G. Liu, Nanoscale thermoplasmonic welding. iScience **25**, 104422 (2022)

# Chapter 2
# Fundamental Principles of Thermoplasmonics

Thermoplasmonic physics can be understood from different points of view [1–5]. A quantum-mechanical microscopic description defines plasmons as the collective oscillations of conduction electrons. Within this framework, plasmons manifest as charge-density waves, oscillating against a background of positive charge from atomic nuclei. The plasmonic oscillations are often attenuated by the excitations of electron–hole pairs within the Fermi Sea of conduction electrons [6–9]. This theoretical model is effectively applied to bulk systems, benefiting from translational symmetry to simplify the complex many-body theory, or to metal clusters with a limited number of conduction electrons. However, metal nanoparticles are excessively large for direct calculation of their properties. A classical, macroscopic approach has successfully characterized the plasmon excitations, applicable even to particles as small as a few nanometers in size [3, 10–12]. This section thus relies on the classical approach to understand thermoplasmonics. The basic fundamentals will be introduced from plasmonic theory and thermodynamics to the thermal-induced processes.

## 2.1 Plasmonic Theory

### *2.1.1 Metal Electromagnetism*

Plasmonics examines how electromagnetic waves within the visible spectrum interact with nanomaterials. In the classical framework, plasmonic response is characterized by the metal particle's localized polarization triggered by external electromagnetic fields. Understanding the dielectric behavior of the metal enables the application of Maxwell's equations to forecast plasmon resonances and their resulting optical properties [3, 12, 13]. The most basic approach to predict a metal's dielectric response is



G. Liu, *Thermoplasmonics*, https://doi.org/10.1007/978-981-97-8332-8_2

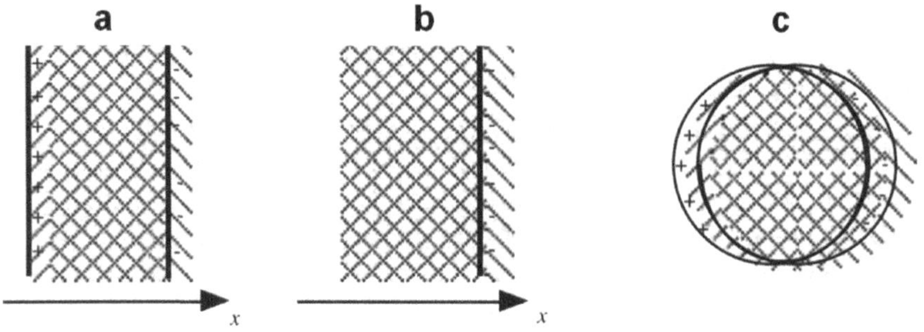


**Fig. 2.1** Electron oscillation modes: the modes of oscillation for the electron gas in **a** a thin film, **b** a metal–vacuum interface, and **c** a nanosphere. Reproduced with permission from Ref. [13] Copyright 2013 Wiley–VCH

the Drude or free electron model. In this model, conduction electrons are conceptualized as a gas of noninteracting, free electrons. These electrons undergo relaxation processes due to collisions with the lattice or other scattering entities (Fig. 2.1). To analyze this behavior, one can employ the electron motion equation within the electron gas framework when exposed to an external electric field **E**:

$$m\ddot{\mathbf{x}} + m\gamma\dot{\mathbf{x}} = -e\mathbf{E} \tag{2.1}$$

here, $m$ represents the effective mass of the electron, and $e$ denotes the elementary charge. The electrons exhibit oscillatory behavior in reaction to the electromagnetic field. Their movement is dampened through collisions, which occur at a characteristic collision frequency $\gamma = 1/\tau$, here where $\tau$ is the mean time between collisions, also known as the relaxation time. **E** signifies the total electric field acting on the electron, including both external fields and the internal field induced by the electron motion. Given the driving field $\mathbf{E}(t) = \mathbf{E}_0 e^{-i\omega t}$, exhibiting harmonic time-dependence, a particular solution to this equation that characterizes the electron's oscillation is presented as $\mathbf{x}(t) = \mathbf{x}_0 e^{-i\omega t}$. Here, the complex amplitude $\mathbf{x}_0$ encapsulates potential phase shifts between the driving field and the oscillatory response, indicated by:

$$\mathbf{x}(t) = \frac{e}{m(\omega^2 + i\gamma\omega)}\mathbf{E}(t) \tag{2.2}$$

Given that electric displacement (**D**) and macroscopic polarization (**P**) are delineated as

$$\mathbf{D} = \varepsilon_0\mathbf{E} + \mathbf{P} = \varepsilon_0\varepsilon\mathbf{E} \tag{2.3}$$

$$\mathbf{P} = -ne\mathbf{x} \tag{2.4}$$

respectively, where $\varepsilon_0$ represents the permittivity of free space, and $\boldsymbol{n}$ denotes the density of conduction electrons, thereby

$$\mathbf{D} = \varepsilon_0 \mathbf{E} - \frac{ne^2}{m(\omega^2 + i\gamma\omega)}\mathbf{E} \tag{2.5}$$

Therefore, the dielectric function of a free electron metal is derived

$$\varepsilon(\omega) = 1 - \frac{\omega_p^2}{\omega^2 + i\gamma\omega} \tag{2.6}$$

where $\omega_p$ represents the plasma frequency of the free electron gas, as defined by

$$\omega_p = \sqrt{\frac{ne^2}{\varepsilon_0 m}} \tag{2.7}$$

This frequency is the characteristic frequency of electron oscillation (or volume plasmons) and is crucial for the optical properties of metals. Light with a frequency greater than $\omega_p$ can transmit through the metal as electrons are too slow to screen it effectively, whereas light is reflected if its frequency is less than $\omega_p$ because the free electrons within the metal are able to shield the electric field effectively.

The dielectric function is a complex physical quantity comprising both real and imaginary components $\varepsilon(\omega) = \varepsilon_1(\omega) + i\varepsilon_2(\omega)$, given by

$$\varepsilon_1(\omega) = \mathrm{Re}\langle\varepsilon(\omega)\rangle = 1 - \frac{\omega_p^2}{\omega^2 + \gamma^2} \tag{2.8}$$

$$\varepsilon_2(\omega) = \mathrm{Im}\langle\varepsilon(\omega)\rangle = \frac{\omega_p^2\gamma}{\omega(\omega^2 + \gamma^2)} \tag{2.9}$$

The real component of the complex dielectric constant relates to the energy stored within the medium, while the imaginary part corresponds to energy dissipation. Another key optical property, the refractive index $n = \sqrt{\varepsilon}$, plays a crucial role in normalizing both the propagation speed ($v = c/n$) and the wavelength of light ($\lambda = \lambda_0/n$) within the medium [14]. If the real part is dominant, it results in a reduced effective propagation speed. Conversely, a significant imaginary component leads to substantial absorption of wave energy within the medium, which in turn causes destructive interference.

Based on common observations, metals block the propagation of electromagnetic waves up to the infrared spectrum and demonstrate significant reflectivity. However, in higher frequencies found in the near-infrared and visible spectrum, there is a tangible increase in the electromagnetic field's penetration, leading to enhanced energy dissipation [3]. As the frequency approaches the ultraviolet range, metals exhibit dielectric properties that allow for the transmission of electromagnetic waves, though the extent of attenuation varies according to the electronic band structure. The frequency-dependent dispersive behavior of this phenomenon can be represented by $\varepsilon(\omega)$. The underlying physics driving this significant optical response is the phase shift of induced currents relative to the driving field, occurring as frequencies near the reciprocal of the characteristic electron relaxation time.

In the visible spectrum, metals generally exhibit high reflectivity due to their plasma frequency $\omega_p$ being in the ultraviolet range and the relatively small value of the damping constant $\gamma$ compared to $\omega$. This leads to a negative real part $\varepsilon_1(\omega) < 0$ and a small imaginary part $\varepsilon_2(\omega)$ of the dielectric function $\varepsilon(\omega)$, allowing for the excitation of plasmon resonance. While the traditional dielectric materials with $\varepsilon_1\omega$ are typically on the order of a few units. It is important to note that interband transitions, mainly occurring at ultraviolet energies, can significantly impact $\varepsilon_2(\omega)$ such as in the case of gold [3]. Considering these interband effects would involve adding a supplementary term to $\varepsilon(\omega)$, which would be merely an additive constant if the transitions occur in the ultraviolet range.

### 2.1.2 Plasmon Resonance Modes

Plasmon resonance involves the rapid oscillation of a free electron cloud within conductive media, prominently observed in metallic nanostructures due to their high electron densities [15, 16]. These structures can be visualized as plasmas of positive ions surrounded by conduction band electrons [17]. During resonance, a metal's optical extinction cross section sharply increases, enhancing near-field light energy concentration at the nanoscale (Fig. 2.2). This significantly affects the optical and photothermal properties of the material, driven principally by two modes: propagating surface plasmon polaritons (SPP) and localized surface plasmon resonance (LSPR). Coupling the plasmon modes amplifies the optical and thermal characteristics of the nanostructures [18–22]. Additionally, adaptable structural designs allow for more complex resonance modes, [23–25] including Fabry-Pérot (F-P) cavity plasmons, gap mode, surface lattice plasmon resonance (SLR), and Fano resonance [21, 26–28]. These collective modes endow plasmonic structures with unique and enhanced properties not achievable by basic modes alone.

LSPR originates from the collective oscillations of conducting electrons confined within a nanoparticle (Fig. 2.2a). These resonances manifest as stationary charge oscillations at optical frequencies along metal boundaries, amplifying electromagnetic fields to sub-wavelength dimensions, which significantly enhances physical processes. The unique optical properties spurred by these resonances have garnered extensive interest from scientific and technological communities. Mie's theory provides an essential framework for analyzing these phenomena, focusing on the interaction between spherical nanoparticles in a dielectric medium and electromagnetic radiation with wavelengths exceeding the particle radius. By solving Maxwell's equations under suitable boundary conditions for extinction and scattering, Mie characterized these interactions as multipolar oscillations. In the quasi-static regime, the

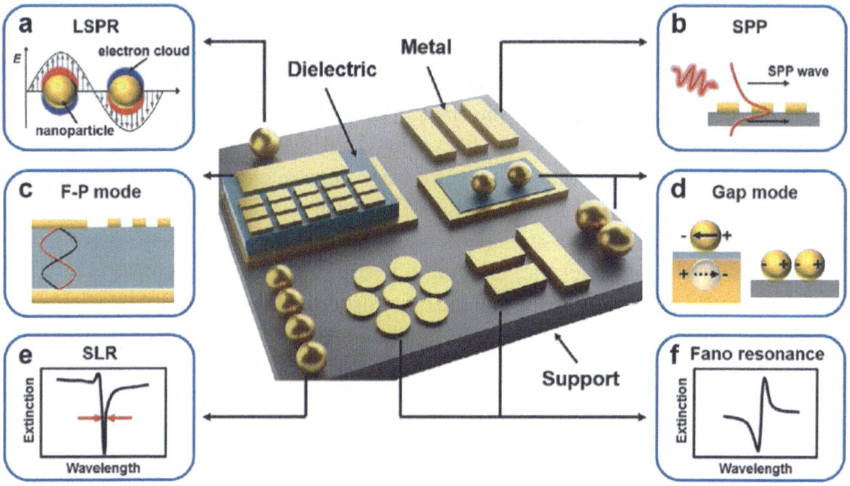


**Fig. 2.2** Plasmon resonances include: **a** Localized surface plasmon resonance (LSPR), **b** Surface plasmon polaritons (SPP), **c** Fabry-Pérot (F-P) cavity mode, **d** Gap mode, **e** Surface lattice resonance (SLR), and **f** Fano resonance. Reproduced with permission from Ref. [28] Copyright 2021 Wiley–VCH

LSPR's influence is described by the particle's polarizability, with the dipole moment determined as follows: [29]

$$\boldsymbol{p} = \alpha \boldsymbol{E}_0 \text{ where } \alpha(\omega) = (1+\psi)V\varepsilon_0 \frac{\varepsilon(\omega) - \varepsilon_m}{\varepsilon(\omega) + \psi\varepsilon_m} \tag{2.10}$$

where $\alpha$ represents the polarizability of the sphere, $\varepsilon_m$ and $\varepsilon(\omega)$ are the permittivity of the medium and the nanostructure, respectively. $\psi$ denotes the shape factor, which has a value of 2 for a sphere. Surfaces with high curvature or sharp features tend to have larger values. For a polarized sphere with a radius of $R$, light absorption is described by the extinction cross section $C_{\text{ext}}$, which is the combination of absorption ($C_{\text{abs}}$) and scattering ($C_{\text{sca}}$) cross sections: [30, 31]

$$C_{\text{ext}} = C_{\text{sca}} + C_{\text{abs}} = \frac{\boldsymbol{k}^4}{6\pi}|\alpha|^2 + \boldsymbol{k}\text{Im}(\alpha) \tag{2.11}$$

$$C_{\text{sca}} = \frac{\boldsymbol{k}^4}{6\pi}|\alpha|^2 = \frac{8\pi}{3}\boldsymbol{k}^4R^6\left|\frac{\varepsilon - \varepsilon_m}{\varepsilon + 2\varepsilon_m}\right|^2 \tag{2.12}$$

$$C_{abs} = \boldsymbol{k}\text{Im}(\alpha) = 4\pi\boldsymbol{k}R^3\text{Im}\left(\frac{\varepsilon - \varepsilon_m}{\varepsilon + 2\varepsilon_m}\right) \tag{2.13}$$

The absorbed energy is primarily dissipated as heat or emitted as photons. However, the quantum yield of photon emission is typically below 1%, [32] suggesting that the majority of absorbed energy converts to heat.

SPPs arise at metal–dielectric interfaces, facilitated by structures like gratings or prisms with variable refractive indices (Fig. 2.2b) [33, 34]. They form when an electromagnetic field with suitable energy causes the free electrons in a metallic thin film to oscillate against the metal's positive ion structure. This collective electron movement, called surface plasmons, couples with an electromagnetic wave in the metal and the adjacent dielectric, creating a hybrid mode known as SPPs. In a flat metal film, SPPs can travel along the surface for tens to hundreds of micrometers, decaying exponentially within hundreds of nanometers in the metal. The propagation length is described as follows: [35]

$$L_{\text{spp}} = \frac{\lambda}{2\pi} \frac{(\varepsilon_1)^2}{\varepsilon_2} \left[ \frac{\varepsilon + \varepsilon_1}{\varepsilon \varepsilon_1} \right]^{\frac{3}{2}} \tag{2.14}$$

Light interacts with the plasmon excitations of a flat, semi-infinite metal surface as achieving simultaneous conservation of energy and momentum is not feasible. Surface plasmons cannot exist independently, especially when retardation effects are notable, but must couple with a transverse electromagnetic wave. To provide the necessary additional momentum, one method involves modifying the planar metal surface with features such as indentations or gratings, which supply "lattice" momentum to ensure momentum conservation. Furthermore, SPPs are often called propagating SPPs due to their decay length being about 50 times longer than that of LSPR. This extended propagation enables the spatial separation between optical excitation and electron collection, reducing background interference.

A Fabry-Pérot (F-P) cavity is an optical resonator formed by two parallel mirrors. It enables standing waves, amplifying select wavelengths via constructive interference and dampening others through destructive interference. Such F-P cavity modes often emerge in stacked plasmonic arrays. Figure 2.2c shows a structure with two metal mirrors and an optical dielectric medium between them, where electromagnetic waves can interfere within this dielectric cavity. The wavelength for the $N$th resonance mode in an F-P cavity, when light is incident perpendicularly, can be calculated as follows: [36, 37]

$$\lambda_N = \frac{2n_d(d + \delta)}{N} \tag{2.15}$$

$n_d$ represents the real part of the dielectric's refractive index, $d$ the thickness between the mirrors, and $\delta$ the effective increase in cavity length caused by reflection phase shifts. The internal electric field and plasmon bandwidth can be tuned using appropriate materials and strategic nanostructure design [21, 23, 28]. Effective plasmonic coupling within the photonic cavity has been achieved by replacing one mirror with a plasmonic element featuring a broad linewidth [38]. This integrated system enhances surface plasmon concentration, significantly reducing radiative damping. Features of this mode include sharply defined spectral properties, high quality factors, and sensitivity to changes in cavity length or refractive indices of the materials involved.

Gap mode plasmon resonance describes the intensification of the electromagnetic field within the nanoscale gap between two closely spaced metallic structures, driven by the excitation of localized surface plasmons at the metal–dielectric interface (Fig. 2.2d). When two plasmonic structures are situated just a few nanometers apart, their electromagnetic fields overlap and interact through coupling, concentrating the electric field within the gap via constructive interference of the plasmon modes [18]. The intensity and wavelength of this resonance are highly sensitive to the gap size; a narrower gap increases coupling, boosts field intensity, and shifts the resonance peak toward longer wavelengths. This mode features electromagnetic field strength several orders of magnitude greater than that of the incoming waves. Both the spectral position and intensity of the resonance respond markedly to variations in the refractive index of the surrounding medium, with different sensitivities depending on each nanoparticle's side. By modifying the particles' shape, size, material, and the surrounding dielectric, it is possible to finely tune the gap resonance across a broad frequency range from visible to infrared. In particle-film setups, an image dipole forms in the film, creating countercharge distributions. These antisymmetric oscillating dipoles strongly confine electromagnetic energy to volumes substantially below the diffraction limit, enhancing applications in nanooptics and nanophotonics.

SLRs occur in periodic arrays of plasmonic particles (Fig. 2.2e). These resonances occur when the diffraction from the periodic lattice combines with the localized surface plasmon resonances of individual nanoparticles. Arranging plasmonic structures in periodic arrays can trigger strong near-field and far-field coupling, resulting in the SLRs. SLRs show a combination of narrow spectral characteristics and large spatial coherence of the electromagnetic fields. The collective resonance of SLRs can suppress radiative damping [21]. The dipole approximation model can be employed to analyze the nature of SLRs, where each plasmonic element is assumed to act as a dipole. For an infinite plasmonic array, the effective polarizability ($\alpha_{\text{eff}}$) of each particle and the extinction cross section ($C_{\text{ext}}$) produced by one particle in the array can be described as [27, 39]

$$\alpha_{\text{eff}} = 1/\left(\frac{1}{\alpha_s} - S\right) \tag{2.16}$$

$$C_{\text{ext}} = 4\pi k \text{Im}(\alpha_{\text{eff}}) \tag{2.17}$$

here $\alpha_s$ symbolizes the polarizability of a solitary metal particle, while $S$ represents the accumulated effect of dipoles from surrounding particles, $k$ the wavenumber of incoming light. SLRs can be tuned by modifying the nanoparticle array's periodicity or by adjusting the size, shape of the nanoparticles, and the refractive index of the adjacent medium, since both $\alpha$ and $S$ depend on these structural factors [28]. However, excessively small particle sizes might hinder SLR development because the dipole contributions may not sufficiently offset particle relaxation effects. Conversely, out-of-plane plasmonic resonances elevate SLR quality. In comparison to LSPRs are sensitive to individual particle discrepancies, SLRs display reduced sensitivity to these variations but are more responsive to long-range order and collective dynamics.

Fano resonance arises as a form of resonant scattering, stemming from the interference between a discrete (localized) state and a continuum of states [21, 26]. This phenomenon manifests in complex hybrid nanostructures, characterized by distinctly asymmetric spectral profiles (Fig. 2.2f), where it can occur due to multipole resonance and radiative dipolar coupling, as described by the Fano formula [40]

$$\sigma_E = D^2 \frac{(q+\Omega)^2}{1+\Omega^2} \tag{2.18}$$

where $D^2 = 4\sin 2\delta$ ($\delta$ represents the phase shift of the continuum), $q = \cot\delta$ is the Fano parameter determining the asymmetry degree, and $\Omega = \frac{2(E-E_0)}{\Gamma}$ ($\Gamma$ is the resonance width and $E_0$ is the energy). Fano resonances commonly manifest in symmetry-breaking nanostructures or nanoparticle clusters, exhibiting sharp subradiant and broad super-radiant plasmon modes, often with high-order and dipole feature modes [28, 41]. These resonances are highly sensitive to changes in system parameters like coupling strength or phase relation between interacting modes. Engineering the line shape of Fano resonance can yield a narrow asymmetric resonance profile with sharp spectral contrast, owing to constructive and destructive interference between narrow discrete and broad continuum modes. Additionally, the near-field electric field can experience significant amplification at the resonance frequency.

### *2.1.3 Quality Factors*

Various quality factors and figures of merit can be applied to quantitatively compare the efficiencies of plasmon resonance [11], such as the quality factor $Q_P$, the figure of merit FOM, the Faraday number Fa, and the Joule numbers Jo. Each of these terms serves an important role in evaluating and designing plasmonic structures for specific applications, with their significance tied to the particular phenomena they describe and measure.

The renowned quality factor $Q_P$ is defined as the ratio of the real part of the permittivity to the imaginary part [42].

$$Q_P = -\frac{\varepsilon_1}{\varepsilon_2} \tag{2.19}$$

The negative sign is used to offset the negative value of $\varepsilon_1$ for achieving a positive number. This equation emphasizes two fundamental principles in plasmonics: $|\varepsilon_1|$ should be high, and $|\varepsilon_2|$ should be low. It represents the ratio of stored energy to energy dissipated per cycle by the plasmonic structure, which is a measure of how long the plasmon oscillations can persist before they are damped out. The maximum values of quality factors and their corresponding frequencies are listed in Fig. 2.3

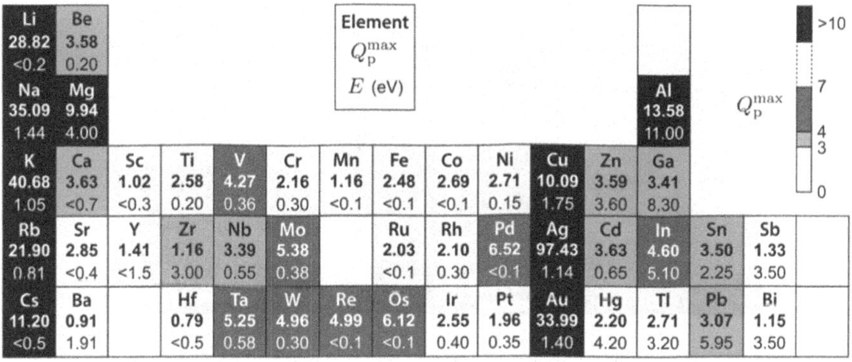


**Fig. 2.3** Quality factors of metals displayed on the Mendeleev's table adapted with permission from Ref. [11, 42] Copyright 2018 Cambridge University Press

for all nongroup-f metals. The values of $\varepsilon_1$ and $\varepsilon_2$ in materials are closely tied to wavelength, making it crucial to select appropriate plasmonic materials to maximize $Q$ within the desired optical range. In the visible spectrum, Ag stands out as a top material for plasmonics thanks to its high quality factor. Nonetheless, Ag is prone to rapid oxidation, leading to significant plasmonic losses. On the other hand, Au is a popular choice in this spectral range, offering outstanding plasmonic performance and favorable chemical stability in normal conditions. When it comes to structures operating in the UV range, Al emerges as a better option due to its high quality factor.

Another metric frequently used to assess plasmonic material is the figure of merit, which characterizes the near-field enhancement by comparing the complex amplitudes [43].

$$\text{FOM} = \frac{\text{Maximum electric near} - \text{field over space}}{\text{Incoming electric field}} = \frac{E_{\text{max}}}{E_0} \tag{2.20}$$

This figure of merit evaluates the efficiency of local electric field enhancement achieved by a plasmonic structure. It quantifies how effectively the plasmonic structure can concentrate the electromagnetic field at its surface or within its near-field region compared to the applied field. Plasmon resonance exhibits a significant ability to confine electromagnetic fields. Plasmonic particles can compress light to nanoscale dimensions, resulting in a remarkable augmentation of the electric field in the vicinity of particle surface [44, 45]. In the dipolar regime, the field enhancement ($\boldsymbol{E}/\boldsymbol{E}_0$) is expressed as: [29]

$$\left(\frac{\boldsymbol{E}}{\boldsymbol{E}_0}\right)_s = \frac{(1+\psi)\varepsilon_m}{\varepsilon + \psi\varepsilon_m} \tag{2.21}$$

The common method involves utilizing near-field simulations such as finite-difference time-domain or finite element method to analyze the electromagnetic fields, offering insights into how effectively the structures amplify the local fields. In

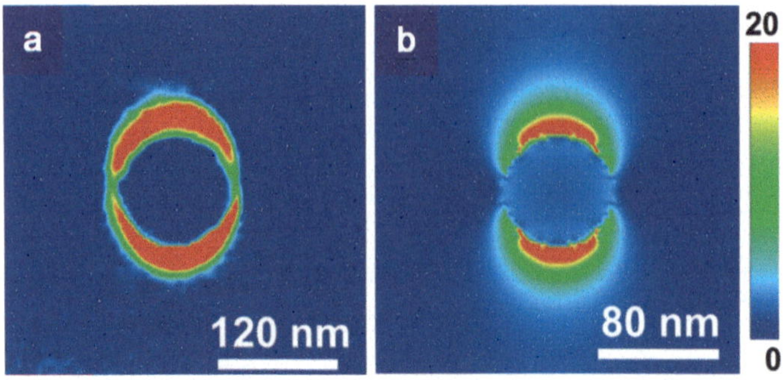


**Fig. 2.4** Electric field enhancements in the vicinity of Ag particle. Experimentally measured (**a**) and FDTD calculated (**b**) electric field intensity maps in the vicinity of nanosphere. Reprinted with permission [48]. Copyright 2010 American Chemical Society

addition, experimental techniques like surface-enhanced Raman scattering or photoemission electron microscopy can be used to map the field enhancements and confirm the simulation results. Experimental measurements and numerical calculations for a 110 nm silver sphere reveal near-field intensity roughly 20 times that of the incident field (Fig. 2.4). Bowtie structures with smaller gaps are projected to exhibit enhancements exceeding 1000 [46]. This metric holds particular significance for applications dependent on robust local field enhancements, including surface-enhanced Raman scattering, microscopy, and spectroscopy, as well as nonlinear optics.

Alternatively, the Faraday number of a plasmon resonance is determined by the ratio of two electric field amplitudes: [47]

$$\mathrm{Fa} = \left| \frac{\text{Maximum electric near} - \text{field over space}}{\text{Incoming electric field}} \right|^2 \tag{2.22}$$

This is different from the FOM definition only by a power of two, including the electric field intensity rather than just the electric field amplitudes. In the quasi-static regime, with a sphere surrounded by a dielectric medium, an exact expression can be derived as: [11, 47]

$$\mathrm{Fa} = 9\left| \frac{\varepsilon}{\varepsilon + \varepsilon_m} \right|^2 \tag{2.23}$$

Note that Fa is independent of the sphere size, as long as the quasi-static approximation holds true, meaning that the sphere is sufficiently small (< 60 nm).

Hereby, Fa represents a figure of merit of the material itself, enabling quantitative comparisons between different materials at any wavelength, not necessarily at the resonance wavelength. The advantage of Fa over $Q_P$ lies in it provides quantitative estimations of the near-field enhancement. It also allows for quantitative comparisons across materials at any wavelength (not just near resonance) and takes account for the effect of the surrounding medium, since it depends not only on $\varepsilon_1$ and $\varepsilon_2$, but also on $\varepsilon_m$.

In parallel with *Fa*, Joule numbers are defined to quantify the capability of plasmonic particle in enhancing the electromagnetic field and producing heat, respectively: [31, 47]

$$\mathrm{Jo} = \frac{e\varepsilon_2}{n_m}|\boldsymbol{E}_{\mathbf{in}}/\boldsymbol{E}_0|^2 \tag{2.24}$$

This equation includes the imaginary part of the material permittivity and the photon energy e in electron volts (eV). In the quasi-static approximation for a sphere, equivalent simple expressions exist:

$$\mathrm{Jo} = \frac{\lambda_{\mathrm{ref}} C_{\mathrm{abs}}}{2\pi V} \tag{2.25}$$

where $V$ represents the nanoparticle volume, and $\lambda_{\mathrm{ref}} = 1240$ nm. This expression is derived by considering the relation $qV = C_{\mathrm{abs}} I$, which emphasizes the connection between Jo and $C_{\mathrm{abs}}$, two quantities relevant to heat generation in plasmonics. The Joule number appears to scale proportionally to the absorption cross section divided by the nanoparticle volume. This observation is intriguing because the conventional method typically involves dividing $C_{\mathrm{abs}}$ by the nanoparticle's projected surface to yield a dimensionless number representing absorption efficiency. However, Joule numbers can be computed for any geometry, indicating that dividing by the nanoparticle volume is more logical, allowing for the extension of Jo's definition to any nanoparticle geometry. This value is crucial for applications where controlled heat generation is required, such as photothermal therapy, thermoplasmonics for material processing, catalysis, or heat-assisted magnetic recording.

## 2.2 Thermodynamics

### *2.2.1 Heat Generation*

Figure 2.5 illustrates a fundamental comprehension of plasmonic heating. At the microscale, plasmon resonances may undergo damping processes, either nonradiatively through the generation of hot carriers via Landau damping or radiatively through photon reemission [49, 50]. The radiance of plasmonic modes determines the branching ratio of these two decay mechanisms. Landau damping is a quantum–mechanical process in which a quantum plasmon is converted into carrier excitation within a timescale of 1–100 femtoseconds. Subsequently, these hot carriers are redistributed through electron–electron scattering occurring at picosecond timescales [15, 33]. Finally, heat is dispersed into the surroundings over an extended timescale spanning picoseconds to nanoseconds, resulting in an overall rise in temperature.

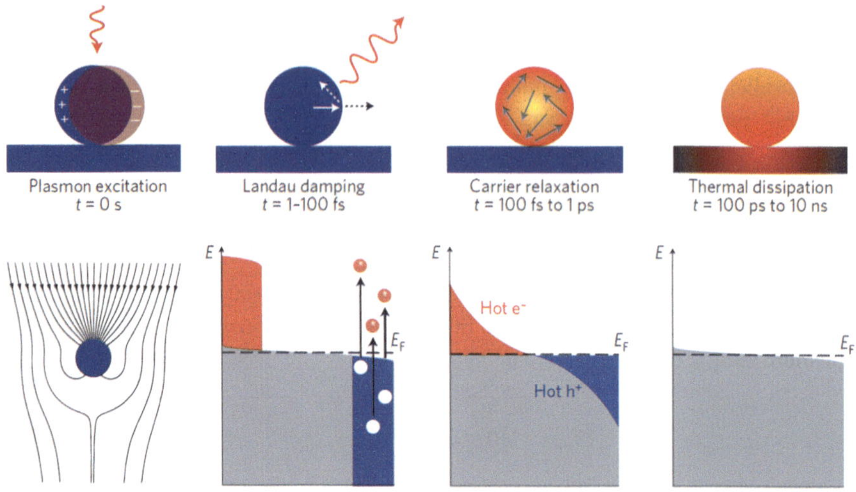


**Fig. 2.5** Key mechanisms driving heat generation involve photoexcitation and relaxation of plasmon resonance, occurring within typical timescales. Reproduced with permission from Ref. [49] Copyright 2015 Nature Publishing Group

From a macroscopic perspective, the free charges within a metal particle oscillate at the frequency of the electric field. This electronic oscillation, akin to an electronic current in a metal, leads to energy dissipation through the Joule effect. According to Poynting's theorem, this energy dissipation can be expressed as: [31, 33, 51]

$$\boldsymbol{q} = \frac{1}{2}\mathrm{Re}\big(\boldsymbol{J}^*(\boldsymbol{r}) \cdot \boldsymbol{E}(\boldsymbol{r})\big) = \frac{1}{2}\varepsilon_0\omega\mathrm{Im}(\varepsilon(\boldsymbol{r}))|\boldsymbol{E}(\boldsymbol{r})|^2 \tag{2.26}$$

This expression involves the loss density of electromagnetic power $\boldsymbol{q}$ and the current density $\boldsymbol{J}$. It is evident that the heat produced in a plasmonic nanostructure is directly related to the square of the electric field within the nanostructure, as well as $\varepsilon_2$. These two factors are conflicting because a high $\varepsilon_2$ results in significant losses, leading to a less enhanced electric field [52]. This consideration is crucial when choosing suitable materials for designing efficient plasmonic nanoheaters.

The total heat power generated by a nanoparticle is the integral of $\boldsymbol{q}$ over the nanoparticle domain:

$$Q = \iiint q\mathrm{d}V = \iiint \frac{1}{2}\mathrm{Re}\big(\boldsymbol{J}^*(\boldsymbol{r})\boldsymbol{E}(\boldsymbol{r})\big)\mathrm{d}V = \iiint \frac{1}{2}\varepsilon_0\omega\mathrm{Im}(\varepsilon(\boldsymbol{r}))|\boldsymbol{E}(\boldsymbol{r})|^2\mathrm{d}V \tag{2.27}$$

The heat generation can also be calculated by use of the absorptive cross section: [53, 54]

$$Q = C_{\mathrm{abs}}I = C_{\mathrm{abs}}\frac{c\varepsilon_0}{2}|\boldsymbol{E}_0|^2 \tag{2.28}$$

$I$ represents the light flux. Dividing the dissipated energy rate by the light intensity yields another expression for the absorptive cross section:

$$C_{\text{abs}} = \frac{Q}{I} = \frac{1}{c\varepsilon_0|\boldsymbol{E}_0|^2} \iiint \varepsilon_0 \omega \text{Im}(\varepsilon(\boldsymbol{r}))|E(\boldsymbol{r})|^2 \text{d}V \tag{2.29}$$

The absorption cross section is a crucial parameter for determining plasmonic heating, and its values are often computed using analytical and numerical methods.

### 2.2.2 Heat Transfer

As temperature increases within a control volume, heat dissipates into the surroundings through conduction, convection, and radiation (see Fig. 2.6) [55–58]. The rate of heat transfer matches the rate of absorbed energy, with the total rate quantified as $\boldsymbol{q} = \sum_i \boldsymbol{q}_i$. Conductive heat transfer typically predominates in thermoplasmonic systems [33, 51, 59].

Heat conduction involves the direct contact and collision of adjacent atoms or molecules, which facilitates energy transfer from warmer to cooler regions in response to a temperature gradient [56]. This phenomenon is governed by Fourier's law, which describes the conductive heat flux due to this gradient:

$$\boldsymbol{q}_{cd}(\boldsymbol{r}, t) = -\kappa \nabla T(\boldsymbol{r}, t) \text{ e.g. } \boldsymbol{q}_{cd} = \kappa \frac{T_1 - T_2}{L} \text{ for a one-dimensional slab} \tag{2.30}$$

Here, $L$ represents the thickness of the medium [53, 60]. The thermal conductivity $\kappa$ indicates the capacity of a medium to conduct heat.

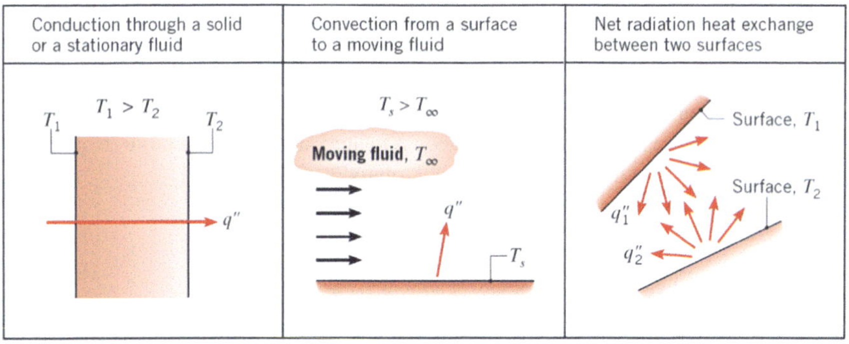


**Fig. 2.6** Conduction, convection, and radiative heat transfer. Reproduced with permission from Ref. [56] Copyright 2017 Wiley–VCH

Heat convection arises from temperature differences between a hot surface and adjacent fluid, also involving phase changes from bubble emission or vapor condensation [56, 61]. In this process, heat first conducts from the surface into a stationary fluid layer, then into the flowing fluid, combining both conduction and thermal transport. Convection can manifest as forced convection, driven by external mechanisms like fans or pumps, or free convection, driven by buoyancy due to temperature-induced density changes. This convective heat transfer is quantified by Newton's cooling law: [56, 62]

$$\boldsymbol{q}_{cv} = h_c(T_s - T_\infty) \tag{2.31}$$

Here, $T_s$ represents the surface temperature, $T_\infty$ the fluid temperature, and $h_c$ the convective heat transfer coefficient, which depends on the fluid's thermodynamic properties, the nature of fluid motion, and surface geometry. The Rayleigh number Ra, representing the ratio of diffusion to convection time scales, is used to determine whether conduction or convection dominates in microscopic systems: [51, 63]

$$\mathrm{Ra} = \frac{g\beta\Delta TR^3}{\nu D} \text{ where } D = \frac{\kappa}{\rho C_p} \tag{2.32}$$

Here, $\beta$ indicates the thermal expansion coefficient, $\nu$ the kinematic viscosity, and $D$ the thermal diffusivity. For nanoscale heating sources with a Ra much less than 1, conduction is predominant, while convection takes precedence in cases involving larger heat sources.

Thermal radiation originates from the movement of charged particles, which emit energy as photons. All surfaces emit thermal radiation, with the emission rate governed by the surface's absolute temperature and properties [64]. The total radiative power emitted by a spherical particle can be calculated using the provided formula: [51]

$$Q_{rd} = 4\pi R^2 \epsilon C_{sb} T^4 \tag{2.33}$$

The particle emissivity is represented by $\epsilon$, and $C_{sb}$ denotes the Stefan–Boltzmann constant. The net heat transfer rate between a heated surface and a large, isothermal surface encompassing the heater can be calculated as: [56, 61, 65]

$$\boldsymbol{q}_{rd} = \epsilon C_{sb}\left(T_s^4 - T_{\mathrm{sur}}^4\right) \tag{2.34}$$

Here, the surrounding temperature is denoted as $T_{\mathrm{sur}}$. In the case of a nanoheater, radiative heat transfer is typically ineffective due to the direct physical contact of the heat source with a high thermal conductivity medium that matches the heater temperature.

### 2.2.3 Temperature Profile

The heat generated inevitably leads to an increase in temperature for both the nanoparticle and its surroundings. By the heat transfer equation, the transient temperature distribution within and around the NP can be described as follows [66]

$$\rho C_p \frac{\partial T(r,t)}{\partial t} = \nabla[\kappa \nabla T(r,t)] + \boldsymbol{q}(r,t) \tag{2.35}$$

This equation relies on two inherent factors that influence heat dissipation: thermal conductivity $\kappa$ and specific heat capacity $C_p$. Thermal conductivity signifies a material's ability to conduct heat, while specific heat capacity represents the energy needed to raise the temperature of a material per unit mass. Mathematical techniques can be introduced to linearize the equation, enabling the calculation of temperature distribution and/or heat flux [51, 56, 60, 67].

For a spherical particle, the temperature rise in both the particle and its surroundings can be approximated as follows: [31, 68]

$$\delta T = \begin{cases} \delta T_P \frac{R}{r} & r > R \\ \delta T_P & r \le R \end{cases} \quad \text{where } \delta T_P = \frac{C_{\text{abs}} I}{4\pi\kappa R} \tag{2.36}$$

Here, $\delta T$ is the temperature increase, $\kappa$ is the thermal conductivity of the medium, and $R$ is the particle radius. Figure 2.7a illustrates the calculated field profiles of gold particles with radii of 100 nm. When the absorption power is 1 μW, the temperature increase reaches approximately 50 K [68]. Inside the particle at the resonance wavelength, a uniform electric field can be observed. However, the retardation effect introduces nonuniformity in energy absorption. Nonetheless, the temperature distribution across the particle remains uniform [69].

Increasing particle concentration reduces interparticle distances due to collective thermal effects [70, 71]. This leads to intensified heating within such particle arrays [72–74]. The cumulative effect and Coulomb interactions among particles further enhance this phenomenon. The total heat varies from the sum of individual particle heats, a result of field screening within particles or plasmon shifts driven by Coulomb interactions. Consequently, the resulting heat depends on interparticle distance, particle number, and arrangement. In an NP array of $N$ elements, the temperature increase of a particle $j$, influenced by neighboring particles i, can be modeled as: [31, 75, 76]

$$\delta T_j = \frac{C_{abs,j} I}{4\pi\kappa R} + \sum_{i \ne j} \frac{C_{abs,i} I}{4\pi\kappa |r_i - r_j|} \tag{2.37}$$

The dominance of the last term, even in sparse particle arrays, may be due to long-range temperature diffusion (Fig. 2.7c, d). [77] The significance of collective effects is evaluated using a dimensionless number, $\xi = p^2/3R\emptyset$, [31, 78] where $p$ is the

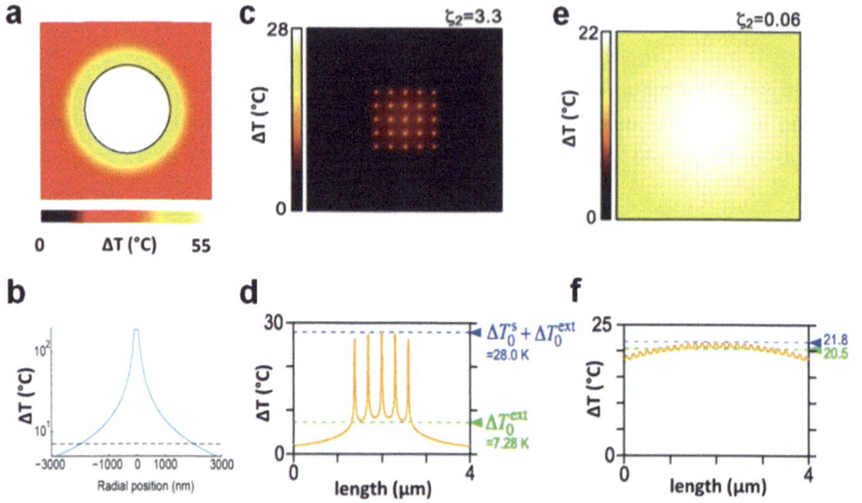


**Fig. 2.7** **a** The temperature increases around an Au NP. Adapted with permission from Ref. [68] Copyright 2010 American Chemical Society. **b** Temperature profile around an Au NP with diameter of 150 nm (adapted from Ref. [69] Copyright 2016 Springer Nature Limited). **c** and **d** Temperature distribution across a finite-size square lattice of NPs. **e** and **f** Temperature distribution across a periodic lattice of NPs. Adapted from Ref. [70] Copyright 2013 American Chemical Society

average interparticle distance, and $\emptyset \sim p\sqrt{N}$ represents the feature size of the particle distribution. Collective effects are dominant when $\xi < 1$ (Fig. 2.7e, f), leading to two distinct regimes: (i) the temperature confinement regime, focused on the temperature distribution around a single hot spot; and (ii) the delocalization regime, characterized by a uniform temperature profile across the medium. Neglecting collective effects can underestimate plasmonic heating in practical applications.

### 2.2.4 *Heating Scheme*

The thermal response is impacted by the heating scheme, such as continuous-wave (CW) illumination and pulsed lasers (nanosecond, picosecond, and femtosecond). The absorption of pulse energy by plasmonic structures occurs in three steps: [79] (1) Electronic absorption: Upon interaction with the laser pulse, the NP free electron gas absorbs part of the pulse energy. These electrons, lighter and more reactive than the ion lattice, quickly reach a Fermi–Dirac distribution within about $\tau_e \sim 100$ [49, 80], creating a nonequilibrium state where the electronic temperature $T_e$ rises while the lattice temperature $T_p$ remains constant. (2) Electron–phonon thermalization: The high-temperature electronic gas cools through electron–phonon interactions $\tau_{e-ph}$ within a characteristic timescale of approximately 1.7 ps for thermalization with the gold lattice ions. This timescale is consistent across NP sizes above 5 nm, with smaller sizes showing variance due to confinement effects. (3) External heat diffusion: Energy diffuses from the nanoparticle to its surroundings over a longer timescale $\tau_{tr} \sim L^2 \frac{C_p \rho}{3\kappa_s}$, resulting in particle cooling and heating of the surrounding medium. This process depends on the system's characteristic size $L$, with the timescale ranging from 100 ps to a few nanoseconds [49]. Understanding these processes involves comparing the timescales of illumination duration $\tau$ and the electron–phonon energy relaxation time $\tau_{e-ph}$, which typically range from $10^{-10}$ to $10^{-12}$ s.

In the CW scenario, including exposure to sunlight or nanosecond-pulsed lasers ($\tau \gg_{e-ph}$), undefined electrons have ample time to transfer their energy to phonons, reaching equilibrium. This process is quasi-equilibrium, allowing the classical heat diffusion equation to apply. In steady-state conditions, the spatial temperature decay from a heat source scales inversely with distance $1/r$, where $r$ is the radial coordinate. Typically, the temperature within nanoparticles remains uniform owing to their significantly higher thermal conductivity than surrounding materials. Thermal resistance between the particle and its surrounding medium does not affect the surrounding medium's temperature in steady state, impacting only the internal temperature of the particle. In transient states, the temperature dynamics of a plasmonic structure do not follow exponential behavior; thus, exponential laws are inappropriate for modeling these dynamics or determining timescales. A more appropriate description involves the use of the error function $\text{erf}\sqrt{\tau/t}$, which varies at a much slower rate than an exponential profile $e^{-t/\tau}$ [81].

Pulsed illumination, spanning femtoseconds to nanoseconds, effectively heats metal nanoparticles. When employing sub-picosecond and femtosecond-pulsed lasers ($\tau \ll \tau_{e-ph}$), the absorption of light by electrons completes before notable electron–phonon scattering occurs. During these brief pulses, this results in heated electrons while the lattice temperature remains unchanged [82]. Since the electronic heat capacity is significantly lower (1–2 orders of magnitude) than that of phonons, it leads to much higher transient electronic temperatures relative to the lattice temperatures, which could vary from 100 to 1000 K depending on pump fluence and pulse duration. Conversely, the lattice temperature increases more gradually, reaching comparatively lower peaks. This rapid thermoplasmonic behavior has been explored using time-resolved pump–probe spectroscopy and advanced modeling approaches [83–85]. Due to the nonequilibrium nature of the system, the traditional thermal model falls short, necessitating the adoption of the two-temperature model that distinctly accounts for the disparate transient temperatures of electrons and phonons.

Compared to CW illumination, pulsed illumination induces a broader spectrum of effects, such as rapid temperature and pressure fluctuations [86], heat confinement [87], acoustic wave generation [88, 89], vibration modes [90, 91], bubble formation [92–94], modification of nanoparticle shapes, melting, and extreme thermodynamic conditions [95–97]. Another important consideration is the spatial temperature distribution near the nanoparticle (Fig. 2.8a). While the temperature follows a $1/r$ relationship under CW illumination, pulsed illumination achieves more localized temperature elevation around the nanoparticle. In the case of a spherical NP, the

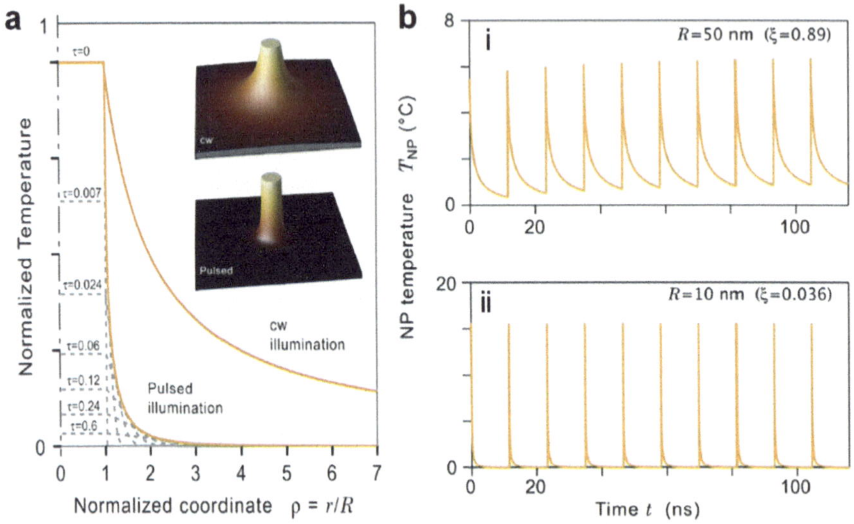


**Fig. 2.8** **a** Radial temperature profiles. Under pulsed illumination, profiles for different normalized times $=_{tr}$ are shown (dashed lines), along with the corresponding temperature envelope. **b** Temporal evolution of the temperature in a spherical gold nanoparticle illuminated by a pulse train at 86 MHz and average irradiance of 1 mW/μm$^2$: (i) $R = 50$ nm; (ii) $R = 10$ nm. Adapted from Ref. [81] Copyright 2012 WILEY–VCH

temperature profile follows a $1/r^3$ relationship for r ≫ R. Additionally, the thermodynamic behavior is affected by the pulsation frequency $f$. If $f$ is too high, the NP may not have sufficient time to fully cool down between consecutive pulses, resulting in a scenario where the NP remains constantly hot (Fig. 2.8b). In practical operation, this state is achieved when $f \geq 1/\tau_{tr}$ [81, 98]. To differentiate between the two scenarios, we can introduce a dimensionless parameter denoted as $\xi = f\tau_{tr}$. When $\xi$ is much smaller than 1, the temperature confinement follows a $1/r^3$ relationship. Conversely, it presents an extended $1/r$ pattern.

## 2.3 Thermal-Induced Physics

### *2.3.1 Thermooptical Effects*

The primary effect of thermoplasmonics is the heating of particles, which alters their optical properties such as absorption and scattering efficiencies through nonlinear thermooptical effects. These effects become pronounced at high temperatures or under intense illumination, influencing optical responses via changes in the damping factor in $\varepsilon_2(\omega)$ [99]. As temperature rises, increased rates of electron–electron and electron–phonon scattering enhance damping losses, leading to bleaching of the

LSP band and broadening of the spectral envelope. This nonlinear thermooptical phenomenon has been harnessed to control heat dissipation in plasmonic nanorods [52].

Additionally, the temperature increase spreads through the surrounding medium, causing refractive index changes, quantifiable by the coefficient *dn/dT*: [31, 100]

$$\delta n = \frac{dn}{dT}\delta T \tag{2.38}$$

Coefficients for gases and solids are generally negligible, whereas liquids like water exhibit significant values. Water refractive index shows considerable nonlinear variation with both temperature (up to 100 °C) and wavelength. Although these coefficients are small, they are substantial enough to produce detectable plasmonic effects, such as altered plasmon resonance, thermal lensing, and increased scattering [101–106].

Thermal emission disperses electromagnetic energy spatially and temporally, [28, 107] releasing broadband photons omnidirectionally into the far-field. The spectral density of these photons correlates with the emitter's temperature as defined by Planck's law [108, 109].

$$I_{\lambda,b}(\lambda, T) = \frac{2hc^2}{\lambda^5\left[\exp(hc/\lambda k_B T) - 1\right]} \tag{2.39}$$

$h = 6.626 \times 10^{-34}$ J is Planck's constant. According to Kirchhoff's law, the emissivity of a material is equal to its absorptivity [64].

$$\varepsilon_\lambda = \alpha_\lambda < 1 \tag{2.40}$$

Metallic structures can be utilized for thermal emission by taking advantage of the enhanced energy absorption at the plasmon resonance. When particle assemblies are in close proximity, a collective thermal effect occurs, leading to enhanced heat flux radiated by the structure [110, 111]

$$Q_{rd} = 4\pi \int_0^\infty \epsilon_\lambda(\lambda) I_{\lambda,b}(\lambda, T)\mathrm{d}\lambda = 4\pi \int_0^\infty \varepsilon_\lambda(\lambda)\frac{2hc^2 C_{\mathrm{geo}}}{\lambda^5\left[\exp(hc/\lambda k_B T) - 1\right]}\mathrm{d}\lambda \tag{2.41}$$

here $C_{geo}$ represents the projected area. However, this expression may result in emissivity values greater than unity. A more accurate formalism utilizing the absorption cross section can be expressed as [31].

$$Q_{rd} = 4\pi \int_0^\infty \frac{2hc^2 C_{abs}(\lambda)}{\lambda^5\left[\exp(hc/\lambda k_B T) - 1\right]}\mathrm{d}\lambda \tag{2.42}$$

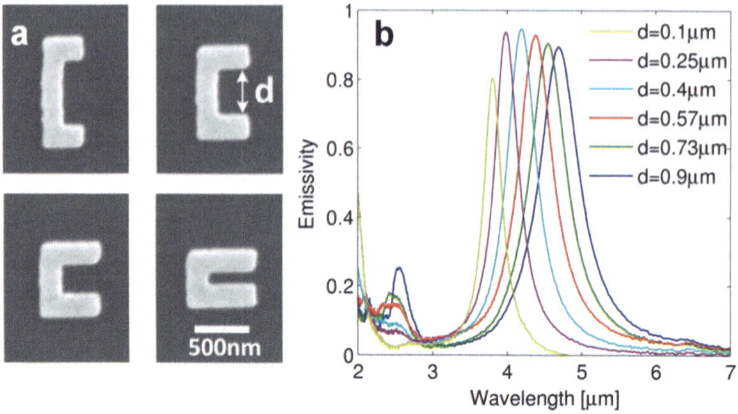


**Fig. 2.9** Thermal radiation and emissivity assessment: **a** SEM images display C-shaped particles with varied gap distances. **b** Measured spectra of y-polarized emissivity. Reproduced with permission from Ref. [111]. Copyright 2017 American Chemical Society

Significant advances are made in tailoring the thermal spectrum, radiation directionality, polarization, and temporal response [112–114]. For example, the emission spectrum of a C-shaped particle emitter displays narrow-band resonance peaks closely associated with its LSPR characteristics (Fig. 2.9) [111, 115]. It reflects that the thermal emission can be tailored by controlling the resonant modes. Typically, thermal emission has a wide range of wavelengths, but it is important to have precise control over the emission in specific narrow bands for various applications. The arrangement of nanorods can significantly influence the thermal radiation by creating a hybrid mode. When the nanorods are closer together, the longitudinal LSPR splits into two modes, resulting in a decrease in the overall thermal emission.

### *2.3.2 Phase Change and Stress Wave*

Thermal waves cannot be produced in physics through harmonic excitation of a nanoparticle since the temperature field follows a diffusion equation. However, phase transitions can occur around plasmonic heaters, causing reshaping or melting of the particle absorber or phase changes in the surrounding medium [31, 66, 116, 117]. For large-scale applications, a solar absorber with high thermal stability is essential. Most absorbers lack the stability needed to operate across broad temperature ranges; for example, gold absorbers must remain below 500 °C to avoid thermal damage. Embedding plasmonic elements in a $SiO_2$ matrix can prevent reshaping [118]. Additionally, all-ceramic plasmonic absorbers composed of TiN and $Al_2O_3$ demonstrate outstanding structural and thermal stability up to 1400 °C [119, 120]. These developments are crucial for creation of thermally stable plasmonic devices suited for extreme environments.

The phase transformations in surrounding materials involve water evaporization and the melting of polymers and ice matrices [66, 121–123]. The liquid–vapor transition is a key phenomenon, typically observed using an array of nanostructures on a substrate [124]. Superheating at the solid/liquid interface must first occur before boiling can start. This metastable state frequently appears around plasmonic particles, where temperatures around gold particles on a glass substrate have been recorded up to 230 °C, and can reach as high as 300 °C if the water is degassed. Bubbles begin to nucleate through vaporization once a critical temperature threshold is reached [125].

Light absorption by metallic structures increases surface temperature, thereby superheating the adjacent liquid and initiating bubble generation and dissipation (Fig. 2.10a) [126]. Bubble dynamics are determined by the balance of pressure forces, fluid inertia, and surface tension as dictated by the Rayleigh-Plesset equation: [97]

$$\rho_L\left(R_b\ddot{R}_b + \frac{3}{2}\dot{R}_b^2\right) + \frac{2\gamma}{R_b} + 4\eta\frac{\dot{R}_b}{R_b} = P_i(t) - P_\infty(t) \tag{2.43}$$

Here, $R_b$ represents the bubble radius, with with $P_i$ and $P_\infty$ indicating internal and external pressures; $\gamma$ is the surface tension, and $\rho_L$ is the liquid's mass density at a distance from the particle. Bubble growth occurs as vapor flows into it due to

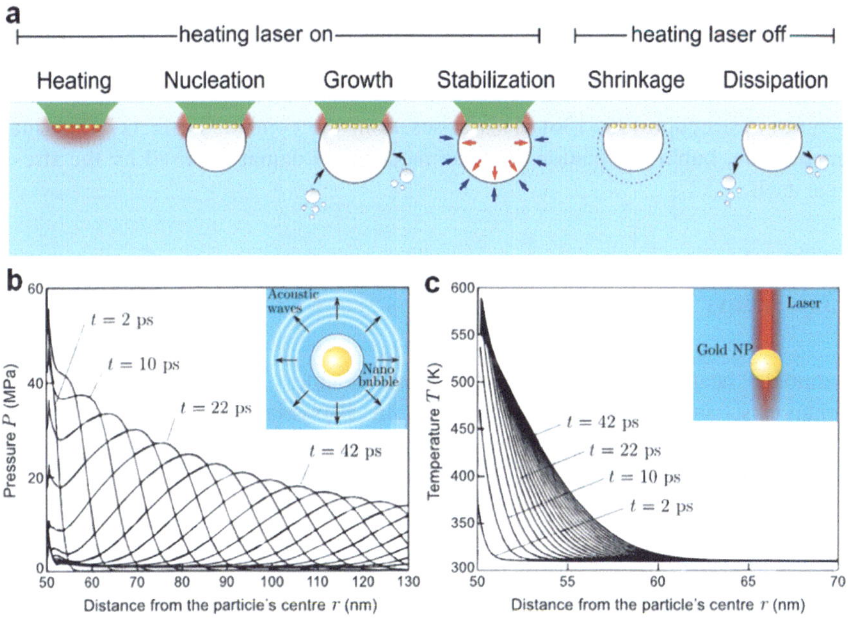


**Fig. 2.10** Bubble dynamics (**a**) and stress wave generation involve (**b**) pressure and (**c**) temperature distribution in water. Simulations used a fluence of 60 J/m$^2$, a pulse duration of 200 fs, and a particle radius of 50 nm. Reproduced with permission [133]. Copyright 2007 Elsevier and [96] Copyright 2021AIP

surrounding water overheating. Upon deactivation of the heating laser, the bubble contracts swiftly as internal pressure falls with vapor condensation. Subsequent vapor condensation into surrounding fluid, mediated by diffusion, notably slows the disappearance of larger bubbles [127].

Bubble formation around a plasmonic structure can significantly raise the structure's temperature, particularly noticeable when bubbles envelop particles in homogeneous media, fully encasing the nanostructure [128–130]. In contrast, for particle arrays on a substrate, the contact area between particles and the substrate critically influences the temperature [124, 131]. The substrate's heat dissipation capacity is essential for preserving the structural integrity.

When plasmonic particles in water reach a local temperature near 300 °C, they quickly transition from liquid to vapor, resembling explosive boiling [86, 132]. This liquid-to-vapor shift is driven by ballistic vapor transport and Kapitza resistance, leading to powerful pressure wave propagation and rapid thermal diffusion at the nanoscale (Fig. 2.10b, c) [86, 131, 133]. This process is often termed shock wave in literature [134, 135]. A shock wave is a high-amplitude pressure wave caused when some molecules move faster than the sound speed in the medium, marked by discontinuities in velocity, density, and pressure. The pressure increase, $\delta P$, can be quantified by using the Grüneisen parameter $\Gamma$, which relates it to the absorbed energy: [31, 96]

$$\delta P = \Gamma Q \tag{2.44}$$

Wave emission from bubble generation provides a novel method for subcellular bacterial destruction [76, 136]. This arises from the combined effects of thermal denaturation, bubble formation, and the mechanical damage caused by the stress wave itself.

### 2.3.3 Mass Transport

Plasmonic heating-induced mass transport includes Brownian motion, thermophoresis, and flow-induced transport. In Brownian motion, nanoparticles randomly collide with fluid molecules driven by thermal fluctuations in the surrounding fluid. This phenomenon is stated by a straightforward relation: [31, 51, 137]

$$\langle \boldsymbol{r}^2 \rangle = 2dD_f t = \frac{dk_B T}{3\pi \eta R} t \tag{2.45}$$

$\langle \boldsymbol{r}^2 \rangle$ symbolizes the mean squared displacement in particle diffusion, where $D_f$ is the diffusion coefficient, $k_B$ the Boltzmann constant, $\eta$ the fluid viscosity, $d$ the dimensionality of motion, and $t$ the time elapsed. The equation highlights how particle displacement depends on its radius, temperature, fluid viscosity, and the timescale considered, assuming thermal equilibrium for particles. Nevertheless, plasmonic heating introduces specific thermal and viscous gradients, leading to "hot Brownian motion," a dynamic state where the adjusted diffusion coefficient reflects either translational or rotational movement.

Active Brownian motion combines self-propelled behavior with conventional Brownian motion, displaying both active swimming and random fluctuations [137, 138]. Active particles absorb environmental energy to maintain nonequilibrium states across varied timescales [139]. On short timescales, particle inertia prevails, leading to ballistic movement. Over long timescales, as inertial effects diminish, particles enter an overdamped regime characterized by a blend of Brownian and active motions or by effective Brownian motion alone. The transition between these regimes depends on the driving force's strength and the rotational diffusion coefficient. Adjusting factors like particle size and aspect ratio can prolong the active motion phase and enhance the effective diffusion coefficient.

Thermal gradients influence particle dynamics at both the particle and fluid scales, each displaying unique characteristics [76, 140]. For particles, thermophoresis—also known as the Ludwig–Soret effect—guides their motion along or away from thermal gradients, causing short-range displacement [31, 76, 141]. This phenomenon facilitates the manipulation of nanoparticles in solutions (Fig. 2.11a). The motion is governed by the equation: [142, 143]

$$v_T = -D_T \nabla T \tag{2.46}$$

Here, $v_T$ represents the solute's thermophoretic velocity, and $D_T$ denotes the thermophoretic mobility. The thermophoretic mobility is influenced by specific interactions at the nanoparticle-solution interface, and its value can vary, being positive or negative depending on the solvent's salinity [144, 145]. For particle suspensions of low concentration, the thermophoretic particle flux $\boldsymbol{J}_{\zeta,\text{thermal}}$ is counterbalanced by the diffusive particle flux $\boldsymbol{J}_{\zeta,\text{diffusive}}$ due to Brownian motion [143].

$$\boldsymbol{J}_{\zeta,\text{net}} = \boldsymbol{J}_{\zeta,\text{thermal}} + \boldsymbol{J}_{\zeta,\text{diffusive}} = -\zeta D_T \nabla T - D_f \nabla T \tag{2.47}$$

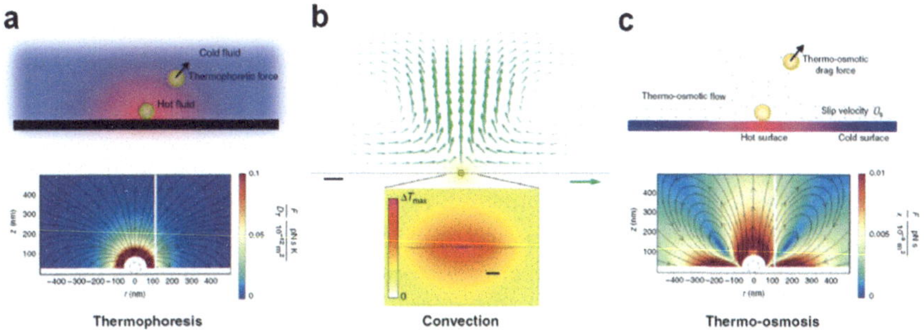


**Fig. 2.11** **a** Thermophoresis, **b** convection, and **c** thermoosmosis. Adapted from Ref. [145] Copyright 2017 American Chemical Society and [146] Copyright 2014 Nature Publishing Group

Here, $\zeta$ represents the local concentration. The significance of thermophoresis and diffusion can be assessed using the Soret coefficient $S_T = D_T/D_f$, which determines the extent of particle depletion or accumulation due to thermophoretic effects.

In the context of fluid dynamics, thermal gradients generate short-range motion through thermoosmotic flow and long-range motion via convection or thermoviscous flow. Convection involves the bulk movement of fluid molecules toward warmer regions' centers, influenced by the heat source's size, temperature increase, and vertical spatial limitations. To enhance convective efficiency, studies have examined thermoplasmonic structures on substrates and indium-tin-oxide coatings on heat-bearing substrates (Fig. 2.11b) [63, 146]. Additionally, temperature gradients alter local viscosity, which has been exploited by directing particle movement with a moving heating spot for thermoviscous motion [147, 148]. Furthermore, these gradients create surface tension gradients that drive thermoosmotic flows capable of reaching speeds over 100 μm/s, thereby clustering particles near the heat sources (Fig. 2.11c).

### 2.3.4 *Chemical Reaction*

Plasmonic chemistry plays a pivotal role in developing clean energy technologies, [149] particularly in solar catalytic reactions that utilize both solar and thermal energies. Understanding the thermal effect is crucial for deciphering the processes involved. [150, 151] Nanoscale thermoplasmonics, by leveraging localized temperature changes, near-field enhancements, and hot-carrier injections, activate chemical reactions [152–154]. The nonequilibrium states of hot carriers, emerging from plasmon decay, can drive reactions through thermal and nonthermal pathways, influencing the kinetics of typically restricted chemical reactions [155, 156].

The reaction encompasses various steps: adsorption of reactants, chemical bond activation, intermediate formation, product conversion, and product desorption [150, 157]. Each of these essential phases requires overcoming specific activation energies ($E_a$) (Fig. 2.12a). As per Arrhenius' law, the reaction rate constant ($K$) is directly related to its molar activation energy: [76, 150]

$$K \propto \exp\left(-\frac{E_a}{RT}\right) \tag{2.48}$$

here $R$ represents the gas constant. This law elucidates how plasmonic heating boosts reaction rates by raising reactant temperatures, effectively lowering the activation energy and enhancing the overall reaction process [158, 159].

Catalytic systems fall into two main categories: pure plasmonic structures and hybrid plasmonic systems [152, 160]. Plasmonic catalysis, involving metallic nanostructures, is well-documented, covering everything from plasmon decay to chemical bond activation and intermediate conversion. In these reactions, hot electrons generally play a more crucial role than thermal effects. The excitation and subsequent decay

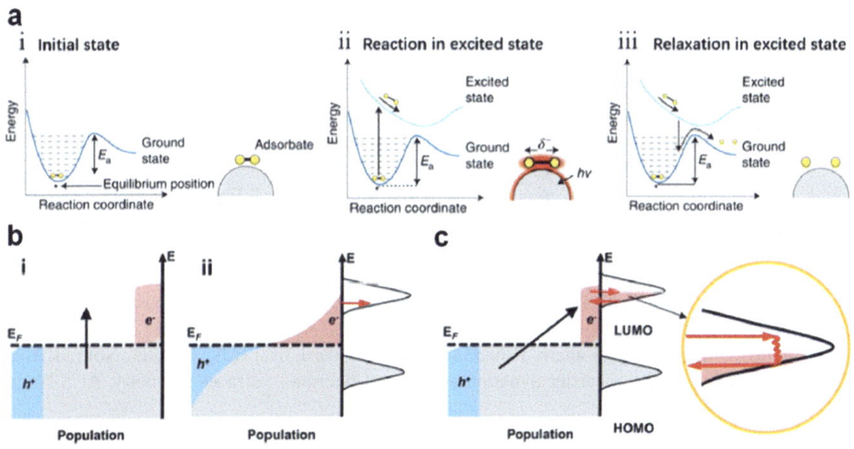


**Fig. 2.12** **a** Plasmon-induced dissociation reaction illustrates (i) the initial state, (ii) the reaction, and (iii) relaxation in the molecule's excited state. Reproduced with permission from Ref. [159] Copyright 2018 Nature Publishing Group. **b** Indirect hot electron transfer from metals to adsorbates involves (i) a nonthermalized distribution of hot carriers and (ii) a Fermi–Dirac distribution of thermalized hot carriers. **c** Direct electron transfer. Reproduced with permission from Ref. [152] Copyright 2017 American Chemical Society

of plasmons facilitate the transfer of hot carriers to adsorbed molecules through both direct and indirect energy transfer mechanisms. These mechanisms result in the formation of adsorbate–metal complex states that enhance redox processes (Fig. 2.12b, c) [152, 157, 161, 162]. Direct electron transfer, although it forms hybridized surface states, has relatively low efficiency due to the minimal transition dipole moments in molecule–metal composites. Indirect electron transfer is more efficient, benefiting from a continuous distribution of hot electrons near the Fermi level, thus serving as the primary excitation pathway.

Hybrid plasmonic systems are promising for solar chemistry, primarily due to their ability to enhance charge separation across the interfacial Schottky barrier [152, 163]. By spatially separating hot electrons from holes at the metal–semiconductor interface, these systems can significantly extend carrier lifetimes [164]. Furthermore, the semiconductor's depletion region effectively sweeps hot electrons away from the interface, reducing charge recombination [165]. This accumulation allows for precise tuning of electron energy levels, while active sites on the semiconductor surface boost catalytic reactions [166, 167]. These hybrid catalysts produce synergistic effects that reduce the activation barriers for chemical reactions. However, the complex interactions between thermoplasmonic and catalytic properties and the mechanisms of energy distribution within these systems require further investigation.

## References


1. G.B. Matthew Pelton, *Introduction to Metal-Nanoparticle Plasmonics*. (Wiley, 2013), p. 275
2. L.V. Besteiro, X.T. Kong, Z. Wang, G. Hartland, A.O. Govorov, Understanding hot-electron generation and plasmon relaxation in metal nanocrystals: quantum and classical mechanisms. ACS Photonics **4**, 2759–2781 (2017)
3. L.D. Sio, *Active Plasmonic Nanomaterials*. (CRC Press, Taylor & Francis Group, LLC, 2016) International Standard Book Number-13: 978-981-4613-4601-4619
4. E. Corte, P.H.C. Camargo, *Plasmonic Catalysis from Fundamentals to Applications*. (Wiley-VCH GmbH, 2021)
5. J. Cunha, T.L. Guo, G. Della Valle, A.N. Koya, R. Proietti Zaccaria, A. Alabastri, Controlling light, heat, and vibrations in plasmonics and phononics. Adv. Opt. Mater. **8** (2020)
6. A.M. Brown, R. Sundararaman, P. Narang, W.A. Goddard 3rd., H.A. Atwater, Nonradiative plasmon decay and hot carrier dynamics: effects of phonons, surfaces geometry. ACS Nano **10**, 957–966 (2016)
7. M.S. Tame, K.R. McEnery, ŞK. Özdemir, J. Lee, S.A. Maier, M.S. Kim, Quantum plasmonics. Nat. Phys. **9**, 329–340 (2013)
8. J. Lee, D.J. Jeon, J.S. Yeo, Quantum plasmonics: energy transport through plasmonic gap. Adv. Mater. **33**, e2006606 (2021)
9. A. Varas, P. García-González, J. Feist, F.J. García-Vidal, A. Rubio, Quantum plasmonics: from Jellium models to ab initio calculations. Nanophotonics **5**, 409–426 (2016)
10. L.V. Besteiro, P. Yu, Z. Wang, A.W. Holleitner, G.V. Hartland, G.P. Wiederrecht, A.O. Govorov, The fast and the furious: ultrafast hot electrons in plasmonic metastructures. Size d structure matter. Nano Today **27**, 120–145 (2019)
11. G. Baffou, *Thermoplasmonics Heating Metal Nanoparticles Using Light*. (Cambridge University Press, 2018) ISBN 978-971-108-41832-41834
12. S.A. Maier, *Plasmonics Fundamentals and Applications*. (Springer Science+Business Media LLC, 2007)
13. S. Enoch, N. Bonod, *Plasmonics From Basics to Advanced Topics*. (Springer Berlin Heidelberg 2012) ISBN 978-973-642-28078-28078
14. S. Jones, *Thermoplasmonic Effects in Microfluidic Systems*, Thesis, Degree of Licentiate of Technology, Chalmers University of Technology, 2019)
15. C. Kuppe, K.R. Rusimova, L. Ohnoutek, D. Slavov, V.K. Valev, "Hot" in plasmonics: temperature-related concepts and applications of metal nanostructures. Adv. Opt. Mater. **8**, 1901166 (2019)
16. L. De Sio, *Active Plasmonic Nanomaterials*. (CRC Press, Taylor & Francis Group, LLC, 2016)
17. Y. Dai, Z. Zhou, A. Ghosh, R.S.K. Mong, A. Kubo, C.B. Huang, H. Petek, Plasmonic topological quasiparticle on the nanometre and femtosecond scales. Nature **588**, 616–619 (2020)
18. N.J. Halas, S. Lal, W.S. Chang, S. Link, P. Nordlander, Plasmons in strongly coupled metallic nanostructures. Chem. Rev. **111**, 3913–3961 (2011)
19. E. Prodan, C. Radloff, N.J. Halas, P. Nordlander, A hybridization model for the plasmon response of complex nanostructures. Science **302**, 419–422 (2003)
20. O. Hess, J.B. Pendry, S.A. Maier, R.F. Oulton, J.M. Hamm, K.L. Tsakmakidis, Active nanoplasmonic metamaterials. Nat. Mater. **11**, 573–584 (2012)
21. B. Wang, P. Yu, W. Wang, X. Zhang, H.C. Kuo, H. Xu, Z.M. Wang, High-Q plasmonic resonances: fundamentals and applications. Adv. Opt. Mater. **9**, 2001520 (2021)
22. L. Feng, P. Huo, Y. Liang, T. Xu, Photonic metamaterial absorbers: morphology engineering and interdisciplinary applications. Adv. Mater. **32**, e1903787 (2020)
23. M. Hertzog, B. Munkhbat, D. Baranov, T. Shegai, K. Borjesson, Enhancing vibrational light-matter coupling strength beyond the molecular concentration limit using plasmonic arrays. Nano Lett. **21**, 1320–1326 (2021)

24. J. Huang, H. Wang, Z. Qi, P. Lu, D. Zhang, B. Zhang, Z. He, H. Wang, Multifunctional metal-oxide nanocomposite thin film with plasmonic Au nanopillars embedded in magnetic La0.67$Sr_{0.33}MnO_3$ Matrix. Nano Lett. **21**, 1032–1039 (2021)
25. A. Teulle, M. Bosman, C. Girard, K.L. Gurunatha, M. Li, S. Mann, E. Dujardin, Multimodal plasmonics in fused colloidal networks. Nat. Mater. **14**, 87–94 (2015)
26. B. Luk'yanchuk, N.I. Zheludev, S.A. Maier, N.J. Halas, P. Nordlander, H. Giessen, C.T. Chong, The Fano resonance in plasmonic nanostructures and metamaterials. Nat. Mater. **9**, 707–715 (2010)
27. C. Cherqui, M.R. Bourgeois, D. Wang, G.C. Schatz, Plasmonic surface lattice resonances: theory and computation. Acc. Chem. Res. **52**, 2548–2558 (2019)
28. K. Yang, X. Yao, B. Liu, B. Ren, Metallic plasmonic array structures: principles, fabrications, properties, and applications. Adv. Mater. e2007988 (2021)
29. J.G. Smith, J.A. Faucheaux, P.K. Jain, Plasmon resonances for solar energy harvesting: a mechanistic outlook. Nano Today **10**, 67–80 (2015)
30. D. Liu, C. Xue, Plasmonic coupling architectures for enhanced photocatalysis, Adv. Mater. e2005738 (2021)
31. G. Baffou, *Thermoplasmonics Heating Metal Nanoparticles Using Light*. (Cambridge University Press, 2017)
32. J. Chen, Z. Ye, F. Yang, Y. Yin, Plasmonic nanostructures for photothermal conversion. Small Sci. **1**, 2000055 (2021)
33. J. Cunha, T.L. Guo, G. Della Valle, A.N. Koya, R. Proietti Zaccaria, A. Alabastri, Controlling light, heat, and vibrations in plasmonics and phononics. Adv. Opt. Mater. **8**, 2001225 (2020)
34. Z. Fang, X. Zhu, Plasmonics in nanostructures. Adv. Mater. **25**, 3840–3856 (2013)
35. A. Shalabney, I. Abdulhalim, Sensitivity-enhancement methods for surface plasmon sensors. Laser Photonics Rev. **5**, 571–606 (2011)
36. R. Ameling, H. Giessen, Cavity plasmonics: large normal mode splitting of electric and magnetic particle plasmons induced by a photonic microcavity. Nano Lett. **10**, 4394–4398 (2010)
37. R. Ameling, H. Giessen, Microcavity plasmonics: strong coupling of photonic cavities and plasmons. Laser Photonics Rev. **7**, 141–169 (2012)
38. T. Xu, W. Zhu, H. Wang, C. Zhang, P.B. Deotare, A. Agrawal, H.J. Lezec, Surface plasmon polariton laser based on a metallic trench Fabry-Perot resonator. Sci. Adv. **3**, e1700909 (2017)
39. V.G. Kravets, A.V. Kabashin, W.L. Barnes, A.N. Grigorenko, Plasmonic surface lattice resonances: a review of properties and applications. Chem. Rev. **118**, 5912–5951 (2018)
40. M.F. Limonov, M.V. Rybin, A.N. Poddubny, Y.S. Kivshar, Fano resonances in photonics. Nat. Photonics **11**, 543–554 (2017)
41. Y.H. Fu, J.B. Zhang, Y.F. Yu, B. Luk'yanchuk, Generating and manipulating higher order fano resonances in dual-disk ring plasmonic nanostructures. ACS Nano **6**, 5130–5137 (2012)
42. M.G. Blaber, M.D. Arnold, M.J. Ford, A review of the optical properties of alloys and intermetallics for plasmonics, Journal of physics. Condensed matter : an Institute of Physics journal **22**, 143201 (2010)
43. P.R. West, S. Ishii, G.V. Naik, N.K. Emani, V.M. Shalaev, A. Boltasseva, Searching for better plasmonic materials. Laser Photonics Rev. **4**, 795–808 (2010)
44. R.B. Claire Deeb, J. Plain, A.-L. Baudrion, S. Jradi, A. Bouhelier, O. Soppera, P.K. Jain, L. Huang, C. Ecoffet, L. Balan, P. Royer, Quantitative analysis of localized surface plasmons based on molecular probing. ACS Nano. **4**, 4579–4586 (2010)
45. O. Nicoletti, F. de la Pena, R.K. Leary, D.J. Holland, C. Ducati, P.A. Midgley, Three-dimensional imaging of localized surface plasmon resonances of metal nanoparticles. Nature **502**, 80–84 (2013)
46. P. Dombi, Z. Pápa, J. Vogelsang, S.V. Yalunin, M. Sivis, G. Herink, S. Schäfer, P. Groß, C. Ropers, C. Lienau, Strong-field nano-optics. Rev. Mod. Phys. **92** (2020)
47. A. Lalisse, G. Tessier, J. Plain, G. Baffou, Quantifying the efficiency of plasmonic materials for near-field enhancement and photothermal conversion. J. Phys. Chem. C **119**, 25518–25528 (2015)

48. C. Deeb, R. Bachelot, J. Plain, A.L. Baudrion, S. Jradi, A. Bouhelier, O. Soppera, P.K. Jain, L. Huang, C. Ecoffet, L. Balan, P. Royer, Quantitative analysis of localized surface plasmons based on molecular probing. ACS Nano **4**, 4579–4586 (2010)
49. M.L. Brongersma, N.J. Halas, P. Nordlander, Plasmon-induced hot carrier science and technology. Nat. Nanotechnol. **10**, 25–34 (2015)
50. J. Cunha, T.L. Guo, A.N. Koya, A. Toma, M. Prato, G. Della Valle, A. Alabastri, R. Proietti Zaccaria, Photoinduced temperature gradients in sub-wavelength plasmonic structures: the thermoplasmonics of nanocones. Adv. Opt. Mater. **8**, 2000568 (2020)
51. S. Jones, Mass Transport via Thermoplasmonics, Thesis Degree of Doctor of Philosophy, Department of Physics, Chalmers University of Technology (2021)
52. B. Yang, C. Li, Z. Wang, Q. Dai, Thermoplasmonics in solar energy conversion: materials, nanostructured designs, and applications. Adv. Mater. **34**, e2107351 (2022)
53. L. Jauffred, A. Samadi, H. Klingberg, P.M. Bendix, L.B. Oddershede, Plasmonic heating of nanostructures. Chem. Rev. **119**, 8087–8130 (2019)
54. G. Baffou, R. Quidant, Thermo-plasmonics: using metallic nanostructures as nano-sources of heat. Laser Photonics Rev. **7**, 171–187 (2013)
55. K.R. Berry, J.R. Dunklin, P.A. Blake, D.K. Roper, Thermal dynamics of plasmonic nanoparticle composites. J. Phys. Chem. C **119**, 10550–10557 (2015)
56. A.S. Lavine, T.L. Bergman, F.P. Incropera, D.P. DeWitt, *Fundamentals of Heat and Mass Transfer*, 8th edn. (Wiley, 2017)
57. A. Yuksel, E.T. Yu, M. Cullinan, J. Murthy, Investigation of heat transfer modes in plasmonic nanoparticles. Int. J. Heat Mass Transf. **156**, 119869 (2020)
58. X. Chen, Y. Chen, M. Yan, M. Qiu, Nanosecond photothermal effects in plasmonic nanostructures. ACS Nano **6**, 2550–2557 (2012)
59. Z. Chen, X. Shan, Y. Guan, S. Wang, J.J. Zhu, N. Tao, Imaging local heating and thermal diffusion of nanomaterials with plasmonic thermal microscopy. ACS Nano **9**, 11574–11581 (2015)
60. M.N. Özisik, D.W. Hahn, *Heat Conduction*, 3rd ed. (Wiley, 2012)
61. Z. Xu, Z. Li, Y. Jiang, G. Xu, M. Zhu, W.C. Law, K.T. Yong, Y. Wang, C. Yang, B. Dong, F. Xing, Recent advances in solar-driven evaporation systems. J. Mater. Chem. A **8**, 25571–25600 (2020)
62. L. Zhu, M. Gao, C.K.N. Peh, G.W. Ho, Solar-driven photothermal nanostructured materials designs and prerequisites for evaporation and catalysis applications. Mater. Horiz. **5**, 323–343 (2018)
63. J.S. Donner, G. Baffou, D. McCloskey, R. Quidant, Plasmon-assisted optofluidics. ACS Nano **5**, 5457–5462 (2011)
64. M.P. Menguc, J.R. Howell, K. Daun, R. Siegel, *Thermal Radiation Heat Transfer*, 7th edn. (CRC Press, 2021)
65. G. Jonsson, D. Tordera, T. Pakizeh, M. Jaysankar, V. Miljkovic, L. Tong, M.P. Jonsson, A. Dmitriev, Solar transparent radiators by optical nanoantennas. Nano Lett. **17**, 6766–6772 (2017)
66. A.O. Govorov, H.H. Richardson, Generating heat with metal nanoparticles. Nano Today **2**, 30–38 (2007)
67. M.I. Tribelsky, A.E. Miroshnichenko, Y.S. Kivshar, B.S. Luk'yanchuk, A.R. Khokhlov, Laser pulse heating of spherical metal particles. Phys. Rev. X **1** (2011)
68. G. Baffou, R. Quidant, F.J. Garcia de Abajo, Nanoscale control of optical heating in complex plasmonic systems. ACS Nano **4**, 709–716 (2010)
69. J.T. Jorgensen, K. Norregaard, P. Tian, P.M. Bendix, A. Kjaer, L.B. Oddershede, Single particle and pet-based platform for identifying optimal plasmonic nano-heaters for photothermal cancer therapy. Sci. Rep. **6**, 30076 (2016)
70. G. Baffou, P. Berto, E. Bermudez Urena, R. Quidant, S. Monneret, J. Polleux, H. Rigneault, Photoinduced heating of nanoparticle arrays. ACS Nano **7**, 6478–6488 (2013)
71. S. Santoro, A.H. Avci, A. Politano, E. Curcio, The advent of thermoplasmonic membrane distillation. Chem. Soc. Rev. **51**, 6087–6125 (2022)

72. V.K. Pustovalov, Light-to-heat conversion and heating of single nanoparticles, their assemblies, and the surrounding medium under laser pulses. RSC Adv. **6**, 81266–81289 (2016)
73. S. Manrique-Bedoya, M. Abdul-Moqueet, P. Lopez, T. Gray, M. Disiena, A. Locker, S. Kwee, L. Tang, R.L. Hood, Y. Feng, N. Large, K.M. Mayer, Multiphysics modeling of plasmonic photothermal heating effects in gold nanoparticles and nanoparticle arrays. J. Phys. Chem. C **124**, 17172–17182 (2020)
74. A. Sepúlveda, D. Boudreau, Neighbors matter: leveraging collective thermoplasmonic effects for smart soft actuators. ACS Appl. Polym. Mater. **6**, 2359–2370 (2024)
75. M.T. Carlson, H.H. Richardson, P.J. Tandler, P. Hernandez, A.O. Govorov, Experimental and theoretical studies of light-to-heat conversion and collective heating effects in metal nanoparticle solutions. Nano Lett., **9**, 1139–1146 (2009)
76. G. Baffou, F. Cichos, R. Quidant, Applications and challenges of thermoplasmonics. Nat. Mater. **19**, 946–958 (2020)
77. M. Mutlu, J.H. Kang, S. Raza, D. Schoen, X. Zheng, P.G. Kik, M.L. Brongersma, Thermoplasmonic ignition of metal nanoparticles. Nano Lett. **18**, 1699–1706 (2018)
78. P.B. Guillaume Baffou, E. Bermúdez Ureña, R. Quidan, S. Monneret, J. Polleux, H. Rigneault, Photoinduced Heating of Nanoparticle Arrays. ACS Nano **7**, 6478–6488 (2013)
79. P. Grua, J.P. Morreeuw, H. Bercegol, G. Jonusauskas, F. Vallée, Electron kinetics and emission for metal nanoparticles exposed to intense laser pulses. Phys. Rev. B **68** (2003)
80. S. Linic, S. Chavez, R. Elias, Flow and extraction of energy and charge carriers in hybrid plasmonic nanostructures. Nat. Mater. **20**, 916–924 (2021)
81. G. Baffou, R. Quidant, Thermo-plasmonics: using metallic nanostructures as nano-sources of heat. Laser Photonics Rev. **7**, 171–187 (2012)
82. R.W. Schoenlein, W.Z. Lin, J.G. Fujimoto, G.L. Eesley, Femtosecond studies of nonequilibrium electronic processes in metals. Phys. Rev. Lett. **58**, 1680–1683 (1987)
83. A. Block, M. Liebel, R. Yu, M. Spector, Y. Sivan, F.J. Garcia de Abajo, N.F. van Hulst, Tracking ultrafast hot-electron diffusion in space and time by ultrafast thermomodulation microscopy. Sci. Adv. **5**, eaav8965 (2019)
84. M. Da Browski, Y. Dai, H. Petek, Ultrafast photoemission electron microscopy: imaging plasmons in space and time. Chem. Rev. **120**, 6247–6287 (2020)
85. P. Samolis, M.K. Hong, R. Rajagopal, M.Y. Sander, S. Erramilli, O. Narayan, Heat transport in photothermal microscopy: newton vs fourier. J. Phys. Chem. C **128**, 961–967 (2024)
86. S.A. Lindley, Q. An, W.A. Goddard 3rd., J.K. Cooper, Spatiotemporal temperature and pressure in thermoplasmonic gold nanosphere-water systems. ACS Nano **15**, 6276–6288 (2021)
87. A. Sanchot, G. Baffou, R. Marty, A. Arbouet, R. Quidant, C. Girard, E. Dujardin, Plasmonic nanoparticle networks for light and heat concentration. ACS Nano **6**, 3434–3440 (2012)
88. F. Magaletti, L. Marino, C.M. Casciola, Shock wave formation in the collapse of a vapor nanobubble. Phys. Rev. Lett. **114**, 064501 (2015)
89. S.Y. Emelianov, P.C. Li, M. O'Donnell, Photoacoustics for molecular imaging and therapy. Phys. Today **62**, 34–39 (2009)
90. J. Rieser, M.A. Ciampini, H. Rudolph, N. Kiesel, K. Hornberger, B.A. Stickler, M. Aspelmeyer, U. Delic, Tunable light-induced dipole-dipole interaction between optically levitated nanoparticles. Science **377**, 987–990 (2022)
91. S.A. Mezzasalma, J. Kruse, S. Merkens, E. Lopez, A. Seifert, R. Morandotti, M. Grzelczak, Light-driven self-oscillation of thermoplasmonic nanocolloids. Adv. Mater. **35**, e2302987 (2023)
92. A.V. Mahulkar, P.S. Bapat, A.B. Pandit, F.M. Lewis, Steam bubble cavitation. AIChE J. **54**, 1711–1724 (2008)
93. S. Moon, Q. Zhang, Z. Xu, D. Huang, S. Kim, J. Schiffbauer, E. Lee, T. Luo, Plasmonic nanobubbles—a perspective. J. Phys. Chem. C **125**, 25357–25368 (2021)
94. A.S. Kostyukov, I.L. Isaev, A.E. Ershov, V.S. Gerasimov, S.P. Polyutov, S.V. Karpov, Part I. Nanobubbles in pulsed laser fields for anticancer therapy: in search of adequate models and simulation approaches. J. Phys. D: Appl. Phys. **55**, 175401 (2022)

95. S. Merabia, S. Shenogin, L. Joly, P. Keblinski, J.L. Barrat, Heat transfer from nanoparticles: a corresponding state analysis. Proc. Natl. Acad. Sci. U.S.A. **106**, 15113–15118 (2009)
96. J. Lombard, J. Lam, F. Detcheverry, T. Biben, S. Merabia, Strong and fast rising pressure waves emitted by plasmonic vapor nanobubbles. Phys. Rev. Res. **3** (2021)
97. J. Lombard, T. Biben, S. Merabia, Kinetics of nanobubble generation around overheated nanoparticles. Phys. Rev. Lett. **112**, 105701 (2014)
98. G. Baffou, H. Rigneault, Femtosecond-pulsed optical heating of gold nanoparticles. Phys. Rev. B **84** (2011)
99. A. Alabastri, A. Toma, M. Malerba, F. De Angelis, R. Proietti Zaccaria, High temperature nanoplasmonics: the key role of nonlinear effects. ACS Photonics **2**, 115–120 (2014)
100. H. Lee, S. Im, C. Lee, H. Lee, S.W. Chu, A.H. Ho, D. Kim, Probing temperature-induced plasmonic nonlinearity: unveiling opto-thermal effects on light absorption and near-field enhancement. Nano Lett. **24**, 3598–3605 (2024)
101. C. Zhao, Y. Liu, Y. Zhao, N. Fang, T.J. Huang, A reconfigurable plasmofluidic lens. Nat. Commun. **4**, 2305 (2013)
102. E. Haque, M. Anwar Hossain, Y. Namihira, F. Ahmed, Microchannel-based plasmonic refractive index sensor for low refractive index detection. Appl. Opt. **58**, 1547 (2019)
103. D. Madaan, A. Kapoor, V.K. Sharma, Ultrahigh sensitivity plasmonic refractive-index sensor for aqueous environment. IEEE Photonics Technol. Lett. **30**, 149–152 (2018)
104. J.S. Donner, J. Morales-Dalmau, I. Alda, R. Marty, R. Quidant, Fast and transparent adaptive lens based on plasmonic heating. ACS Photonics **2**, 355–360 (2015)
105. Y. Shen, J. Zhou, T. Liu, Y. Tao, R. Jiang, M. Liu, G. Xiao, J. Zhu, Z.K. Zhou, X. Wang, C. Jin, J. Wang, Plasmonic gold mushroom arrays with refractive index sensing figures of merit approaching the theoretical limit. Nat. Commun. **4**, 2381 (2013)
106. M.J. Horton, O.S. Ojambati, R. Chikkaraddy, W.M. Deacon, N. Kongsuwan, A. Demetriadou, O. Hess, J.J. Baumberg, Nanoscopy through a plasmonic nanolens. Proc. Natl. Acad. Sci. **117**, 2275–2281 (2020)
107. P.T. David Boyer, A. Maali, B. Lounis, M. Orrit, Photothermal imaging of nanometer-sized metal particles among scatterers. Science **297**, 1160–1163 (2002)
108. D.G. Baranov, Y. Xiao, I.A. Nechepurenko, A. Krasnok, A. Alù, M.A. Kats, Nanophotonic engineering of far-field thermal emitters. Nat. Mater. **18**, 920–930 (2019)
109. S. Fan, Thermal photonics and energy applications. Joule **1**, 264–273 (2017)
110. H. Kallel, R. Carminati, K. Joulain, Temperature of a nanoparticle above a substrate under radiative heating and cooling. Phys. Rev. B **95** (2017)
111. B. Liu, W. Gong, B. Yu, P. Li, S. Shen, Perfect thermal emission by nanoscale transmission line resonators. Nano Lett. **17**, 666–672 (2017)
112. J.W. Cho, S.J. Park, S.J. Park, Y.B. Kim, Y.J. Moon, S.K. Kim, Cooling metals via gap plasmon resonance. Nano Lett. **21**, 3974–3980 (2021)
113. J. Li, B. Yu, S. Shen, Scale law of far-field thermal radiation from plasmonic metasurfaces. Phys. Rev. Lett. **124** (2020)
114. A.C. Overvig, S.A. Mann, A. Alù, Thermal metasurfaces: complete emission control by combining local and nonlocal light-matter interactions. Phys. Rev. X **11** (2021)
115. J. Li, J. Wuenschell, Z. Li, S. Bera, K. Liu, R. Tang, H. Du, P.R. Ohodnicki, S. Shen, Fiber coupled near-field thermoplasmonic emission from gold nanorods at 1100 K. Small **17**, 2007274 (2021)
116. A.B. Taylor, A.M. Siddiquee, J.W.M. Chon, Below melting point photothermal reshaping of single gold nanorods driven by surface diffusion. ACS Nano **8**, 12071–12079 (2014)
117. G. Gonzalez-Rubio, A. Guerrero-Martinez, L.M. Liz-Marzan, Reshaping, fragmentation, and assembly of gold nanoparticles assisted by pulse lasers. Acc. Chem. Res. **49**, 678–686 (2016)
118. D. Dalacu, L. Martinu, Temperature dependence of the surface plasmon resonance of Au/$SiO_2$ nanocomposite films. Appl. Phys. Lett. **77**, 4283–4285 (2000)
119. T. Krekeler, S.S. Rout, G.V. Krishnamurthy, M. Störmer, M. Arya, A. Ganguly, D.S. Sutherland, S.I. Bozhevolnyi, M. Ritter, K. Pedersen, A.Y. Petrov, M. Eich, M. Chirumamilla, Unprecedented thermal stability of plasmonic titanium nitride films up to 1400 °C. Adv. Opt. Mater. 2100323 (2021)

120. Y. Li, C. Lin, Z. Wu, Z. Chen, C. Chi, F. Cao, D. Mei, H. Yan, C.Y. Tso, C.Y.H. Chao, B. Huang, Solution-processed all-ceramic plasmonic metamaterials for efficient solar-thermal conversion over 100–727 °C. Adv. Mater. **33**, e2005074 (2021)
121. H.H. Richardson, Z.N. Hickman, A.O. Govorov, A.C. Thomas, W. Zhang, M.E. Kordesch, Thermooptical properties of gold nanoparticles embedded in Ice: characterization of heat generation and melting. Nano Lett. **6**, 783–788 (2006)
122. F. Meder, G.A. Naselli, A. Sadeghi, B. Mazzolai, Remotely light-powered soft fluidic actuators based on plasmonic-driven phase transitions in elastic constraint. Adv. Mater. **31**, 1905671 (2019)
123. M.S. Zielinski, J.W. Choi, T. La Grange, M. Modestino, S.M. Hashemi, Y. Pu, S. Birkhold, J.A. Hubbell, D. Psaltis, Hollow mesoporous plasmonic nanoshells for enhanced solar vapor generation. Nano Lett. **16**, 2159–2167 (2016)
124. S. Jones, D. Andren, T.J. Antosiewicz, M. Kall, Ultrafast modulation of thermoplasmonic nanobubbles in water. Nano Lett. **19**, 8294–8302 (2019)
125. K. Metwally, S. Mensah, G. Baffou, Fluence Threshold for photothermal bubble generation using plasmonic nanoparticles. J. Phys. Chem. C **119**, 28586–28596 (2015)
126. S. Jones, D. Andren, T.J. Antosiewicz, A. Stilgoe, H. Rubinsztein-Dunlop, M. Kall, Strong transient flows generated by Thermoplasmonic bubble nucleation. ACS Nano (2020)
127. Y. Wang, M.E. Zaytsev, G. Lajoinie, H.L. The, J.C.T. Eijkel, A. van den Berg, M. Versluis, B.M. Weckhuysen, X. Zhang, H.J.W. Zandvliet, D. Lohse, Giant and explosive plasmonic bubbles by delayed nucleation. Proc. Natl. Acad. Sci. **115**, 7676–7681 (2018)
128. Z. Fang, Y.R. Zhen, O. Neumann, A. Polman, F.J. Garcia de Abajo, P. Nordlander, N.J. Halas, Evolution of light-induced vapor generation at a liquid-immersed metallic nanoparticle. Nano Lett. **13**, 1736–1742 (2013)
129. O. Neumann, A.S. Urban, J. Day, S. Lal, P. Nordlander, N.J. Halas, Solar vapor generation enabled by nanoparticles. ACS Nano **7**, 42–49 (2013)
130. Y. Wang, M.E. Zaytsev, H.L. The, J.C. Eijkel, H.J. Zandvliet, X. Zhang, D. Lohse, Vapor and gas-bubble growth dynamics around laser-irradiated, water-immersed plasmonic nanoparticles. ACS Nano **11**, 2045–2051 (2017)
131. Y. Zhang, A. Prosperetti, Dynamics, heat and mass transfer of a plasmonic bubble on a solid surface. Int. J. Heat Mass Transf. **167**, 120814 (2021)
132. L. Hou, M. Yorulmaz, N.R. Verhart, M. Orrit, Explosive formation and dynamics of vapor nanobubbles around a continuously heated gold nanosphere. New J. Phys. **17**, 013050 (2015)
133. A.N. Volkov, C. Sevilla, L.V. Zhigilei, Numerical modeling of short pulse laser interaction with Au nanoparticle surrounded by water. Appl. Surf. Sci. **253**, 6394–6399 (2007)
134. F. Magaletti, L. Marino, C.M. Casciola, Shock wave formation in the collapse of a vapor nanobubble. Phys. Rev. Lett. **114** (2015)
135. A. Dagallier, E. Boulais, C. Boutopoulos, R. Lachaine, M. Meunier, Multiscale modeling of plasmonic enhanced energy transfer and cavitation around laser-excited nanoparticles. Nanoscale **9**, 3023–3032 (2017)
136. L. Wang, Y. Feng, K. Wang, G. Liu, Solar water sterilization enabled by photothermal nanomaterials. Nano Energy **87**, 106158 (2021)
137. H. Sipova-Jungova, D. Andren, S. Jones, M. Kall, Nanoscale inorganic motors driven by light: principles, realizations, and opportunities. Chem. Rev. **120**, 269–287 (2020)
138. I. Buttinoni, G. Volpe, F. Kümmel, G. Volpe, C. Bechinger, Active brownian motion tunable by light. J. Phys. Condens. Matter **24**, 284129 (2012)
139. B. ten Hagen, S. van Teeffelen, H. Löwen, Brownian motion of a self-propelled particle. J. Phys. Condens. Matter **23**, 194119 (2011)
140. B. Ciraulo, J. Garcia-Guirado, I. de Miguel, J. Ortega Arroyo, R. Quidant, Long-range optofluidic control with plasmon heating. Nature communications **12** (2021)
141. L. Lin, X. Peng, M. Wang, L. Scarabelli, Z. Mao, L.M. Liz-Marzan, M.F. Becker, Y. Zheng, Light-directed reversible assembly of plasmonic nanoparticles using plasmon-enhanced thermophoresis. ACS Nano **10**, 9659–9668 (2016)

142. Z. Chen, P.S. Kollipara, H. Ding, A. Pughazhendi, Y. Zheng, Liquid optothermoelectrics: fundamentals and applications. Langmuir ACS J. Surf. Colloids **37**, 1315–1336 (2021)
143. S. Liu, L. Lin, H.B. Sun, Opto-thermophoretic manipulation. ACS Nano **15**, 5925–5943 (2021)
144. L. Lin, M. Wang, X. Peng, E.N. Lissek, Z. Mao, L. Scarabelli, E. Adkins, S. Coskun, H.E. Unalan, B.A. Korgel, L.M. Liz-Marzan, E.L. Florin, Y. Zheng, Opto-thermoelectric nanotweezers. Nat. Photonics **12**, 195–201 (2018)
145. J. Gargiulo, T. Brick, I.L. Violi, F.C. Herrera, T. Shibanuma, P. Albella, F.G. Requejo, E. Cortés, S.A. Maier, F.D. Stefani, Understanding and reducing photothermal forces for the fabrication of Au nanoparticle dimers by optical printing. Nano Lett. **17**, 5747–5755 (2017)
146. B.J. Roxworthy, A.M. Bhuiya, S.P. Vanka, K.C. Toussaint Jr., Understanding and controlling plasmon-induced convection. Nat. Commun. **5**, 3173 (2014)
147. M. Mittasch, P. Gross, M. Nestler, A.W. Fritsch, C. Iserman, M. Kar, M. Munder, A. Voigt, S. Alberti, S.W. Grill, M. Kreysing, Non-invasive perturbations of intracellular flow reveal physical principles of cell organization. Nat. Cell Biol. **20**, 344–351 (2018)
148. F.C. Martin Fränzl, *Hydrodynamic Manipulation of Nano-objects by Thermo-Osmotic Flows*. Cornell University (2021)
149. S. Linic, P. Christopher, D.B. Ingram, Plasmonic-metal nanostructures for efficient conversion of solar to chemical energy. Nat. Mater. **10**, 911–921 (2011)
150. Z.J. Wang, H. Song, H. Liu, J. Ye, Coupling of solar energy and thermal energy for carbon dioxide reduction: status and prospects. Angew. Chem. **59**, 8016–8035 (2020)
151. G. Baffou, I. Bordacchini, A. Baldi, R. Quidant, Simple experimental procedures to distinguish photothermal from hot-carrier processes in plasmonics. Light Sci. Appl. 9 (2020)
152. Y. Zhang, S. He, W. Guo, Y. Hu, J. Huang, J.R. Mulcahy, W.D. Wei, Surface-plasmon-driven hot electron photochemistry. Chem. Rev. **118**, 2927–2954 (2018)
153. C. Zhan, Q.X. Wang, J. Yi, L. Chen, D.Y. Wu, Y. Wang, Z.X. Xie, M. Moskovits, Z.Q. Tian, Plasmonic nanoreactors regulating selective oxidation by energetic electrons and nanoconfined thermal fields. Sci. Adv. **7**, eabf0962 (2021)
154. J. Li, S.K. Cushing, F. Meng, T.R. Senty, A.D. Bristow, N. Wu, Plasmon-induced resonance energy transfer for solar energy conversion. Nat. Photonics **9**, 601–607 (2015)
155. E. Cortés, Activating plasmonic chemistry. Science **362**, 28–29 (2018)
156. L. Zhou, D.F. Swearer, C. Zhang, H. Robatjazi, H. Zhao, L. Henderson, L. Dong, P. Christopher, E.A. Carter, P. Nordlander, N.J. Halas, Quantifying hot carrier and thermal contributions in plasmonic photocatalysis. Science **362**, 69–72 (2018)
157. Z. Zhang, C. Zhang, H. Zheng, H. Xu, Plasmon-driven catalysis on molecules and nanomaterials. Acc. Chem. Res. **52**, 2506–2515 (2019)
158. C. Zhan, M. Moskovits, Z.Q. Tian, Recent progress and prospects in plasmon-mediated chemical reaction. Matter **3**, 42–56 (2020)
159. U. Aslam, V.G. Rao, S. Chavez, S. Linic, Catalytic conversion of solar to chemical energy on plasmonic metal nanostructures. Nat. Catal. **1**, 656–665 (2018)
160. S. Linic, U. Aslam, C. Boerigter, M. Morabito, Photochemical transformations on plasmonic metal nanoparticles. Nat. Mater. **14**, 567–576 (2015)
161. J.M.P. Martirez, J.L. Bao, E.A. Carter, First-principles Insights into plasmon-induced catalysis. Annu. Rev. Phys. Chem. **72**, 99–119 (2021)
162. E. Kazuma, J. Jung, H. Ueba, M. Trenary, Y. Kim, Real-space and real-time observation of a plasmon-induced chemical reaction of a single molecule. Science **360**, 521–526 (2018)
163. J.W. Hong, D.H. Wi, S.U. Lee, S.W. Han, Metal-semiconductor heteronanocrystals with desired configurations for plasmonic photocatalysis. J. Am. Chem. Soc. **138**, 15766–15773 (2016)
164. J.A. Tomko, E.L. Runnerstrom, Y.S. Wang, W. Chu, J.R. Nolen, D.H. Olson, K.P. Kelley, A. Cleri, J. Nordlander, J.D. Caldwell, O.V. Prezhdo, J.P. Maria, P.E. Hopkins, Long-lived modulation of plasmonic absorption by ballistic thermal injection. Nat. Nanotechnol. **16**, 47–51 (2021)
165. K. Wu, J. Chen, J.R. McBride, T. Lian, Efficient hot-electron transfer by a plasmon-induced interfacial charge-transfer transition. Science **349**, 632–635 (2015)

166. M. Wang, M. Ye, J. Iocozzia, C. Lin, Z. Lin, Plasmon-mediated solar energy conversion via photocatalysis in noble metal/semiconductor composites. Adv Sci (Weinh) **3**, 1600024 (2016)
167. P. Christopher, M. Moskovits, Hot charge carrier transmission from plasmonic nanostructures. Annu. Rev. Phys. Chem. **68**, 379–398 (2017)

# Chapter 3
# Thermoplasmonic Materials

Material selection, design, and fabrication are critical aspects in thermoplasmonics. Here, we focus on fundamental strategies for selecting suitable plasmonic materials, tuning surface plasmon responses through single nanoparticle designs involving size and shape, and leveraging coupling effects in NP assemblies with carefully designed matrices or substrates. Once the selection and design are established, the common synthesizing methods for novel nanostructures are further presented to achieve optimal thermoplasmonic properties.

## 3.1 Material Selection

Plasmon resonance is an intrinsic property of materials, with the dielectric function playing a crucial role in governing optical excitation and the propagation length of surface plasmon polaritons. The optical response of a conductive material to incident light is determined by its complex polarizability or dielectric permittivity [1, 2]. The latter describes the bulk properties of the material, including contributions from electronic band transitions, vibrational mode excitations, and the movement of free carriers [3]. Carrier mobility and interband losses are also key factors influencing the optical properties. Lower carrier mobilities lead to higher material losses due to scattering effects caused by defects in the solid [4, 5]. Interband losses occur when electrons absorb incident photons and transition to higher energy levels, representing the primary loss mechanism in materials at optical frequencies [6]. Together with losses from intraband transitions, these mechanisms contribute to the intrinsic material loss, which is well represented by the imaginary part of the dielectric function $\varepsilon_2$. The peak position of localized surface plasmon resonance is determined by the real part of the dielectric function. A larger real part indicates effective screening, resulting in a redshift of the plasmon peak. Conversely, a larger imaginary part corresponds to higher energy losses [7, 8]. However, this parameter is intrinsic and cannot



G. Liu, *Thermoplasmonics*, https://doi.org/10.1007/978-981-97-8332-8_3

be easily modified once established. Consequently, selecting the appropriate material is a crucial initial step in designing plasmon resonance [9, 10].

Figure 3.1 presents the basic optical properties of plasmonic materials based on three key parameters: carrier concentration, carrier mobility, and interband losses [4]. To effectively excite surface plasmons (SPs), the carrier concentration must be sufficiently high to yield a negative real part of the dielectric permittivity. Additionally, $\varepsilon_2$ should be low enough to ensure well-defined SPs [11]. Although higher intrinsic material losses might seem to contribute predominantly to heat generation, it has been shown that the SP-induced heating capacity is generally proportional to $1/\varepsilon_2$ [12]. Consequently, ideal thermoplasmonic materials should exhibit high carrier concentration and mobility, coupled with relatively low losses, positioning them on the left end of the plot. Unfortunately, such an optimal material is currently unavailable. In this context, the advantages and disadvantages of various plasmonic materials, including metals, metallic derivatives, heavily doped semiconductors, and 2D materials, along with their selection criteria, are discussed as follows.

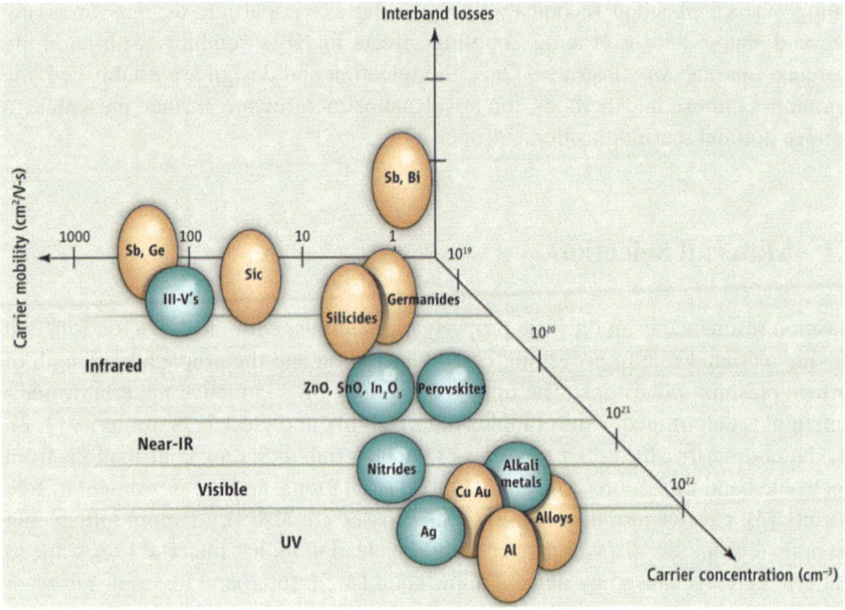


**Fig. 3.1** Critical parameters include carrier concentration, mobility, and interband losses. Spherical bubbles indicate materials with low interband losses; elliptical bubbles signify those with higher interband losses in the respective spectrum. Reproduced with permission [4]. Copyright 2011, AAAS

### 3.1.1 Metal Nanostructures

Gold and silver are the predominant materials in thermoplasmonics because of their high density of conduction electrons and relatively low Ohmic losses. These characteristics result in significant surface plasmon responses across a broad spectrum, ranging from the visible to the near-infrared region (see Fig. 3.1) [13–15]. However, gold and silver experience significant interband transitions that introduce additional losses at optical frequencies, thereby limiting their tenability [10]. Moreover, silver is prone to oxidation or sulfidation in atmospheric conditions, substantially degrading its plasmonic properties. Additionally, both metals are rare and expensive, posing challenges for large-scale applications. In contrast, copper and aluminum are considered promising alternatives due to their abundance and cost-effectiveness [9, 16]. Copper exhibits plasmonic responses similar to those of gold, but it suffers from chemical instability when exposed to atmospheric conditions [17]. In contrast, aluminum shows strong surface plasmons in the ultraviolet range and exhibits the most significant near-field enhancement and highest photothermal efficiency, supporting high-quality SPs across the UV–visible-infrared spectrum [12, 18, 19]. Although aluminum is prone to rapid oxidation, it typically forms a thick oxide layer that serves as an effective passivation coating. However, a notable limitation of Al is the broader linewidths of their SPs in the visible spectrum, primarily due to substantial radiative losses from interband transitions [20, 21].

Refractory metals such as tungsten (W), tantalum (Ta), and molybdenum (Mo) have garnered significant interest because of their high melting points. The ability to maintain chemical stability at temperatures exceeding 2000 °C makes them appealing for high-temperature applications [22, 23]. Nevertheless, these refractory metals exhibit surface plasmon resonances in the near-infrared region and are characterized by relatively high losses. Metals like cobalt, chromium, nickel, palladium (Pd), platinum (Pt), titanium, and others have also been investigated for supporting surface plasmons in the ultraviolet range [1, 24, 25]. However, most of the metals are not particularly practical for thermoplasmonics, with the notable exceptions of Pd and Pt. The catalytic properties of Pd and Pt make them valuable in related fields of nanochemistry. Furthermore, although alkali metals possess appealing plasmonic properties, their extremely high chemical reactivity excludes them from practical applications [10]. Beyond the high losses and poor chemical stability that plague many metals, additional challenges in thermoplasmonics include difficulties in nanofabrication and issues with surface roughness, which can introduce further losses. A significant drawback of metals is their optical properties not easily tuned by external electric, optical, or thermal fields due to their high density of conduction electrons [9, 10].

### 3.1.2 Metallic Derivatives/Semiconductor Crystals

The high carrier concentration in pure metals contributes to their losses. A common strategy to mitigate this issue is to blend them with less conductive metals or nonmetals, resulting in the formation of alloys, intermetallics, or metallic compounds such as silicides, germanides, borides, carbides, nitrides, and oxides [4]. This can be accomplished by alloying or combining two or more elements, allowing for the precise tuning of surface plasmon responses by the adjustment of each proportion [26, 27]. Alloying or doping modifications can lead to remarkable alterations in a material's complex optical constants. However, successful instances of such modifications through changes in chemical composition, structural phase, or carrier density have been relatively rare. For instance, alloying Ag with Au allows for precise tuning of their optical behavior and exhibits a notable redshift as the Au fraction increases [28, 29]. The alloy properties are not simply a linear combination of the optical constants of its components. Instead, a more nuanced understanding, incorporating the shifts in the optical gap and Fermi level is necessary. Generally, the optical properties of an alloy tend to be inferior to those of the best individual metal constituents, and intermetallics with larger unit cells often experience higher optical losses [28].

There are also metallic compounds to consider [30–33]. Notably, transition metal nitrides like titanium nitride (TiN) and zirconium nitride (ZrN) have emerged as promising alternatives due to their exceptional refractory and plasmonic properties [22, 34]. These materials boast remarkable thermal stability, withstanding temperatures above 2900 °C. They support surface plasmons in the visible and longer wavelength regions and provide a good impedance match with air. However, they also exhibit high inherent losses due to larger carrier relaxation rates and interband transitions in optical ranges, making them less effective as near-field enhancers but more efficient as nanoheaters [1, 35, 36]. For example, TiN has garnered interest for its potential in high-temperature applications [37–39]. ZrN and hafnium nitride (HfN) also exhibit plasmonic characteristics at longer wavelengths. With their high melting points, these materials are excellent candidates to replace noble metals in high-temperature scenarios, enabling the tuning of plasmonic resonance through compositional adjustments [32, 40, 41].

Semiconductors are often translucent in the infrared region but opaque in the visible spectrum. To enhance their absorption, doping is used to increase free charge carrier concentrations [42]. Doping causes several modifications to the electronic band structure of semiconductors: i) the creation of intraband energy states, ii) the narrowing of the bandgap, and iii) the formation of impurity bands in degenerately doped transition metal semiconductors. These changes collectively broaden the absorption range of the materials to some extent (Fig. 3.2) [42–44]. For example, plasmon resonances have been tuned in sub-stoichiometric $MoO_{3-x}$ nanobelts with high oxygen vacancy, showing flexible plasmon absorption in a broad range of wavelength from 200 to 2500 nm. This is because the oxygen vacancy induces electron transitions between defect levels and other energy levels [45]. The second effect

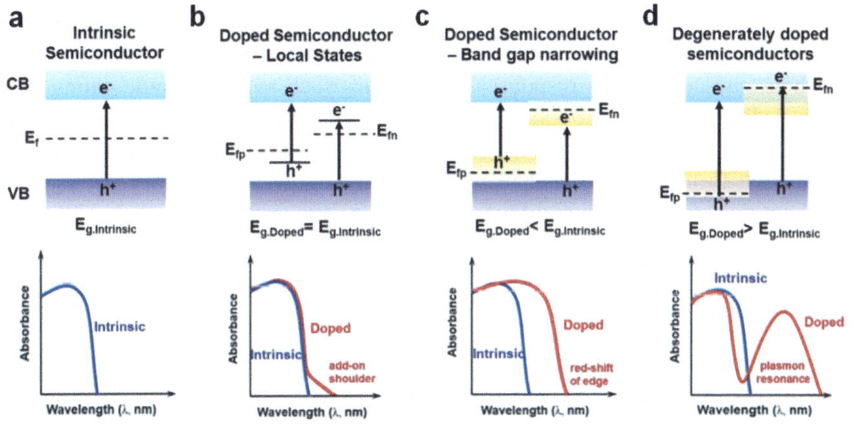


**Fig. 3.2** Bandgap engineering of semiconductors: The band structure and optical absorption profiles of **a** an intrinsic semiconductor, **b** intraband energy states induced by doping, **c** bandgap narrowing caused by doping, and **d** localized surface plasmon resonances induced by degenerate doping. Reprinted with permission from Ref. [43]. Copyright 2019 Royal Society of Chemistry

is exemplified by narrowing of the $Ti_2O_3$ nanoparticle bandgap, enabling absorption across the entire solar spectrum [46]. Recently, it has been found that heavily doped semiconductors exhibit strong localized surface plasmon resonances in the NIR region. Increasing the doping concentration not only enhances the intensity of the LSPR band but also causes blue shifts [47, 48]. Generally, both n-type ($In_2O_3$, CdO, ZnO, $MoO_{3-x}$, $WO_{3-x}$) and p-type ($Cu_{2-x}S$, $Cu_{2-x}Se$, $Cu_{2-x}Te$, $Fe_{1-x}S_2$) semiconductors can be heavily doped to achieve high charge-carrier concentrations. Such doping broadens the optical absorption range and enhances photothermal conversion efficiency due to increased nonradiative recombination probabilities in the presence of more free carriers. However, the doping levels in many semiconductors are often limited by factors such as solid solubility limits, the fraction of active dopants, and doping compensation effects [6, 49]. Some challenges still persist, such as the difficulty in controlling surface plasmons in metal oxides due to dopant-dependent phase transformations and ligand exchanges. Additionally, metal chalcogenides often exhibit instability under high temperatures or laser illumination [50].

### *3.1.3 2D Materials*

Atomically thin 2D materials exhibit considerable promise as plasmonic materials, offering distinct advantages over their bulk counterparts [51, 52]. The most studied 2D van der Waals (vdW) materials include graphene [53, 54], transition metal dichalcogenides [55], black phosphorus [56, 57], and various other materials [58, 59].

These 2D materials are highly tunable, can achieve a broad range of doping concentrations, and possess favorable depolarization factors, enabling improved control over their surface plasmon responses. Techniques such as electrostatic gating and chemical doping have been widely employed to enhance their carrier density. For instance, 2D $MoS_2$ can support surface plasmons in the far-infrared and terahertz regimes due to its low carrier density, typically ranging from $10^{12}$ to $10^{14}$ $cm^{-3}$. By $Li^+$ intercalation, ultradoped $MoS_2$ nanoflakes were synthesized, shifting the SP resonance into the visible and near-UV ranges [60]. However, tuning SPs across such a broad wavelength range is challenging for most 2D vdW materials. Even for graphene, its SPs can be electrically, chemically, or optically tuned mostly within ranges of THz-mid infrared (MIR), and into the NIR in some cases [59]. Additionally, the large-scale fabrication of single-crystalline 2D vdW materials remains a significant challenge, limiting their practical applications [61].

Another notable family of 2D materials is MXenes, which are 2D carbides and nitrides of transition metals derived by selectively etching strongly bonded layered solids [62, 63]. These materials exhibit high metallic conductivity ($n_e \sim 10^{22}$ $cm^{-3}$) and offer chemically tunable plasmonic properties. Their optical characteristics are influenced by the type and structure of the M and X sites, as well as the stoichiometry of surface terminations, offering significant potential for high tenability [63, 64]. Different MXenes can support surface plasmons across the entire visible to near-infrared spectral region, with strong absorption in the UV range due to interband transitions [65, 66]. For example, $Ti_3C_2$, a prominent representative of MXenes, demonstrates superior solar absorption compared to carbon nanotubes, achieving an internal photothermal conversion efficiency close to 100% [67]. Thanks to their strong light absorption and heating effects, 2D MXenes are promising candidates for efficiently harnessing light energy and converting it into thermal energy [68, 69].

To compare the material efficiencies, dimensionless parameters *Fa* and *Jo* are often derived to quantify their plasmonic performance, specifically their ability to enhance electromagnetic fields and generate heat [12, 31]. Figure 3.3 shows the specific values of *Fa* and *Jo* for nanospheroids at plasmon resonance. The data indicate that deviating from a spherical shape can enhance both the optical near field and photothermal conversion efficiency while causing a red shift in the plasmon resonance. Ag particles exhibit the highest resonance intensity for plasmon excitation [70]. In plasmonic applications, metallic materials remain predominant, with gold being the preferred choice due to its excellent photothermal properties and chemical stability in the presence of oxygen [71–73].

## 3.2 Material Design

The electric fields surrounding particles that generate plasmon oscillations are influenced by the size and shape of particles. Particles with sharp points and high aspect ratios lead to concentrated electric fields, resulting in lower restoring forces and consequently lower resonance frequencies. Additionally, when separate

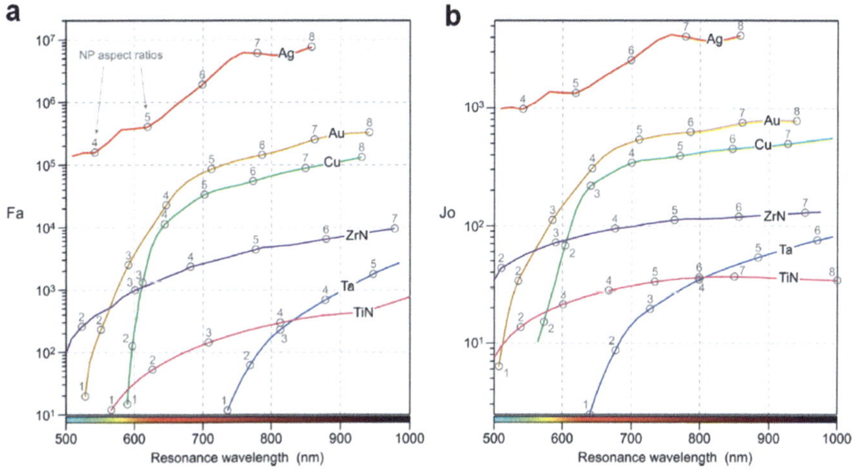


**Fig. 3.3** Dimensionless numbers representing the plasmonic material's ability to enhance the near-field (*Fa*) and generate heat (*Jo*), plotted as functions of the resonance wavelength for ellipsoids with a specific aspect ratio. Reproduced with permission from Ref. [12]. Copyright 2015 American Chemical Society

nanoparticles are brought close together, their plasmons couple, causing the resonance frequency to shift further downward and concentrating the fields into smaller volumes. By designing nanoparticle assemblies, it is possible to tune plasmon resonances to match specific optical frequencies, enabling the confinement of optical fields within three-dimensional spaces as small as a few nanometers. Hereby, in this section, we discuss the size, shape as well as the coupling effect of material design on the plasmon resonance.

### 3.2.1 Size Effect

A primary objective in thermoplasmonics is to maximize photothermal conversion. Equations (2.12) and (2.13) reveal that the scattering cross section becomes increasingly significant with an increase in particle size since it scales with $R^6$ while absorption scales with $R^3$. Consequently, as particle size increases, the scattering cross section progressively dominates the extinction cross section. Generally, increasing the particle size leads to a redshift and broadening of the resonant peaks (Fig. 3.4). This phenomenon arises from retardation effects in the electromagnetic interactions between distant charges, which exhibit nonuniform oscillation states in larger nanoparticles [1]. To optimize energy conversion, the total extinction energy increases with the particle size. However, the dominant mechanism shifts from absorption in smaller NPs to scattering in larger ones (Fig. 3.4a–b), reflecting a competition between radiative and nonradiative energy decay processes

[74]. The broadening of resonant peaks is due to increased radiative damping in larger nanoparticles, leading to a decrease in the amount of energy that can be stored in the localized surface plasmons. Consequently, this reduces the maximum local electric field intensity [74, 75]. Therefore, smaller nanoparticles are more efficient absorbers compared to their larger counterparts, making them potentially more advantageous for thermoplasmonic applications.

For particles smaller than the light wavelength (< 25 nm), absorption is the dominant process over scattering. As the particle size decreases from around 20 nm, the plasmon band initially experiences a blue shift before exhibiting a significant red shift when approaching about 12 nm [76]. Additionally, the plasmon bandwidth widens as particle size decreases, which is inversely proportional to the radius of the particle [77, 78]. Hereby, smaller nanoparticles demonstrate efficient absorption at their resonance wavelength, highlighting their effectiveness in converting light into heat [79, 80]. For larger particles, the plasmon bandwidth undergoes a red shift and broadens with increasing particle size due to retardation effects. This phenomenon occurs because the particle can no longer be approximated as an oscillating dipole. For gold particles, these effects become significant at sizes around 50 nm, with scattering becoming dominant above 90 nm [81]. When particle size exceeds this

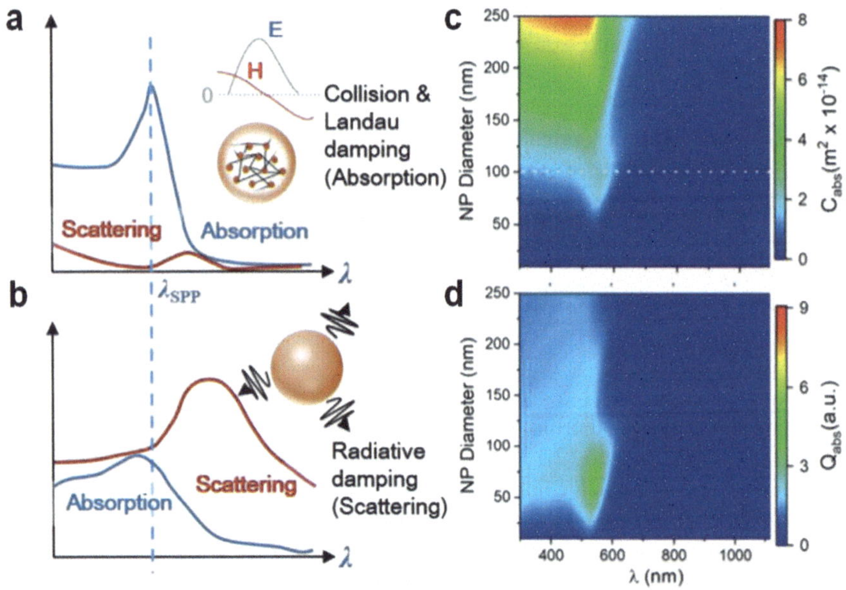


**Fig. 3.4** Damping mechanisms vary with particle size: **a** In smaller particles, absorption-induced nonradiative damping predominates; **b** In larger particles, scattering-induced radiative damping becomes dominant. Reproduced with permission from Ref. [82]. Copyright 2022, Wiley–VCH GmbH. Absorption **c** cross section and **d** efficiency of a single Au particle as functions of the particle radius and wavelength. Reproduced from Ref. [79]. Copyright 2019 American Chemical Society

threshold, retardation effects arise from the electromagnetic interactions between distant charges within the particle [31]. The most widely studied materials are gold and silver nanoparticles, with Ag exhibiting a greater cross section than Au [30]. The shift in absorption wavelength is relatively small compared to the overall bandwidth of resonance. Hence, the peak position is not discussed in relation to the size effect [77].

Experimental studies have focused on gold nanoparticles [79]. The plasmonic heating of a 100 nm gold particle in water has been studied using a near-infrared laser. The observed heating aligns closely with the predicted values derived from dipole approximation [83]. Additionally, Mie theory is extended to evaluate the extinction spectra of core–shell nanostructures [84]. By adjusting the geometric dimensions, tunable plasmon bands and photothermal effects are achieved. The plasmonic band shifts toward longer wavelengths as the shell-to-radius ratio decreases. The maximum observed temperature increase was 110 K for a nanostructure with a core radius of 53 nm and a shell thickness of 8 nm when irradiated at a 900 nm wavelength. This allows the prediction of optimized structures to enhance light-to-heat conversion.

### *3.2.2 Shape Effect*

Adjusting the particle shape is another way to tune plasmon absorption [85]. This tunability arises from either the deviation from dipole oscillation or the coupling between elementary plasmons within a single particle [86–88]. This can be understood from three distinct aspects. Firstly, by controlling particle symmetry, one can manipulate the number of resonances. For example, small gold particles exhibit a single plasmon resonance band at 520 nm, but breaking the symmetry from spherical to cylindrical results in two resonance bands [81]. This asymmetry induces a significant red shift in the plasmon resonance and a substantial increase in resonance amplitude. The two resonance peaks correspond to transverse and longitudinal oscillations with multipolar modes, aligned with the short and long axes of the high aspect ratio nanorods. This variation in electron oscillation arises from different restoring forces along each axis [78, 89]. The band position can also be modified by altering the aspect ratio of different geometries. Figure 3.5a shows the spectra of anisotropic particles with a fixed short axis of 10 nm and a variable long axis ranging from 20 to 80 nm. These particles include ellipsoids, cylinders, nanoprisms, and nanorods. The transverse resonance band (depicted with dashed lines) of all shapes is weak and exhibits a blue shift within the visible-light spectrum. In contrast, the longitudinal plasmon resonance (shown with solid lines) is strong and red shifts into the red or near-infrared regions as the aspect ratio increases. By controlling the aspect ratio, it is possible to flexibly tune the resonance peak positions from the visible to the NIR spectrum [90]. Compared to spherical particles, nanorods are more efficient nanoheaters due to their larger surface area-to-volume ratio, which allows more surface electrons to participate in resonance. Nanorods with an aspect ratio of 3 achieve a maximal heat

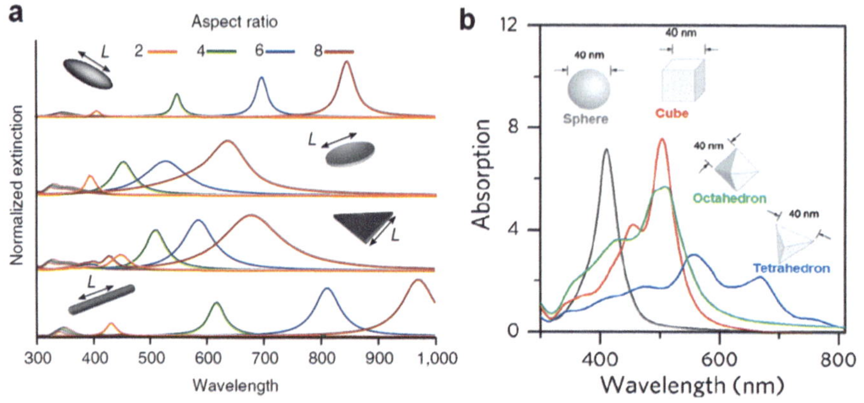


**Fig. 3.5** **a** Spectra of high aspect ratio particles, including ellipsoids, cylindrical disks, triangular nanoprisms, and nanorods. Reproduced with permission from Ref. [90]. Copyright 2014 Nature Publishing Group. **b** The dependence of absorption spectra on particle shape. Adapted with permission from Ref. [80, 93]. Copyright 2006 American Chemical Society

power density that is about four times greater than that of similarly sized spherical particles [91].

The second aspect is controlling the plasmon length where carrier oscillations occur [92]. The plasmon length, a critical parameter for characterizing plasmonic nanostructures, affects not only the resonance peak but also its bandwidth in plasmon resonance. This parameter can be tuned by variations in particle size and shape, allowing for customization of the electron oscillation distance along a plasmon length. Investigations into gold particles with diverse shapes, such as nanocubes, icosohedra, octahedra, decahedra, and truncated bitetrahedra, reveal this by measuring and normalizing their optical spectra against their edge lengths and plasmon lengths. The studies indicate that for dipole modes across a size spectrum of 50–350 nm, edge effects are negligible and exhibit no consistent relationship with side length. Conversely, particles possessing similar plasmon lengths manifest comparable plasmon resonance spectra, regardless of their shape. Hence, the oscillation distance is a crucial factor in describing the shape-dependent plasmonic properties.

The final aspect considers the pronounced tips of structures, such as hollow nanocubes, nanostars, or nanoprisms, facilitate intricate light–matter interactions [94]. This typically results in plasmon resonances spanning a wide range of wavelengths [87, 95–98]. Figure 3.5b illustrates the spectral variations among silver particles of different geometries. Spherical and cubic Ag particles, characterized by their high symmetry, exhibit pronounced plasmonic absorption peaks. In contrast, Ag particles with low-symmetry polyhedral shapes display a wide absorption band that is less intense and includes multiple resonant frequencies [93]. The plasmonic band of the nanocube exhibits a red shift when compared to that of the spherical particle. This transition can be attributed to the accumulation of charges at the corners of the cube, which diminishes the restoring force for electron oscillation [80]. Furthermore, the field intensity around structures with sharp tips can be exceptionally high due to the concentration of polarization charges at these points. The nanostar is a prime example of such structures, which have been employed in photochemical and biomedical applications due to their unique field-enhancing characteristics [79, 99]. Plasmon absorption is highly sensitive to the length of the spikes. An increase in spike length results in a red shift of the plasmon band, whereas the tip curvature or the number of spikes has minimal impact on the absorption peak.

### *3.2.3 Intraparticle Coupling*

The surrounding medium acts as an additional factor in tuning plasmonic coupling [100–102]. Equation (2.7) indicates that the dielectric function of both the constituent materials and the surrounding medium determines the plasmonic characteristics and the corresponding dipole moment. This dielectric coupling is crucial for designing plasmon-coupled resonances. The surrounding dielectric medium can be categorized either as absorbing or nonabsorbing [81]. In the nonabsorbing case, the dielectric function of the medium is characterized by a real dielectric constant, and the induced polarization charges in the environment increase with this constant. Consequently, as the dielectric constant increases from water to glass, the Coulombic restoring force is weakened, resulting in a red shift of the plasmon resonance peak [101].

The absorptive medium features a complex dielectric function, encompassing materials such as polymers, $Al_2O_3$, $SiO_2$, $MoS_2$, graphene, quantum dots, and other metal oxides [7, 81, 103]. In these cases, plasmon resonance is influenced by both the spectral shift due to changes in the real part of the dielectric constant and the shielding effect due to light absorption associated with the imaginary part. The geometry and coupling structure of the materials determine whether the plasmonic resonance will be significantly enhanced or completely suppressed [81]. When the electronic absorption energy of the surrounding medium is far from the resonance energy, an increase in the real part of the dielectric function leads to a red shift of the resonance peak, whereas variations in the imaginary part affect the attenuation of the plasmon resonance [104]. However, in a strong coupling regime, where the absorption energy of the medium is close to the resonance energy, the medium can offset the intrinsic absorption losses in the metal, thereby enhancing the plasmon resonance [7, 77]. This strong coupling may also cause peak splitting, resulting in the creation of hybridized resonance states [105, 106].

The strong intraparticle coupling can significantly alter absorption properties, with core–shell structures and nanomatryoshkas serving as prime examples [107]. The former typically consists of either a dielectric core with a metallic shell or the reverse, while the latter is characterized by a four-layer concentric nanoparticle with alternating dielectric and metallic layers. The plasmonic responses in these nanoparticles can be understood as the coupling or hybridization of fundamental plasmons found in simpler nanostructures. Figure 3.6 presents the localized surface plasmon

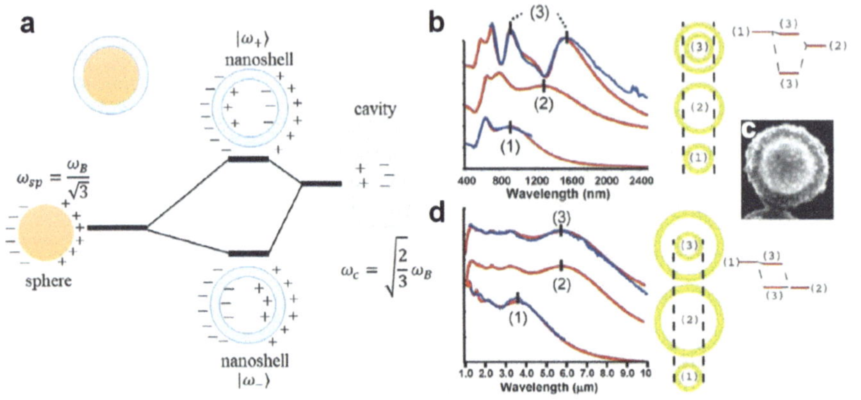


**Fig. 3.6** Intraparticle coupling. **a** An energy-level diagram illustrating plasmon hybridization in metal nanoshells due to the coupling between the spherical core and the cavity. **b**, **d** Experimental (blue) and calculated (red) extinction spectra of elementary nanostructures in the nanomatryoshka for **b** strongly coupled plasmons and **d** uncoupled plasmons. **c** SEM images showing the four-layer concentric nanoshell structure. Reproduced with permission from Ref. [106]. Copyright 2019 AAAS

responses of nanoshells, which arise from the coupling between the plasmon response of a spherical core with relatively constant frequency and that of a nanocavity, both of which are significantly influenced by their geometries. In a four-layer nanomatryoshka composed of silica and gold, the thickness of the intervening dielectric layer determines the coupling intensity between the plasmons of the inner and outer shells. This, in turn, dictates the energy distribution of the hybridized plasmons, as illustrated in Fig. 3.6b, d [106]. This concept of plasmon hybridization provides general principles for designing complex nanostructures with arbitrary shapes.

Various metallic and dielectric materials have been engineered to optimize plasmonic heating [100, 108]. The thermodynamic challenges in this context are complex due to the presence of media with different thermal conductivities. Efficient heat management becomes crucial once it is generated. Microscopically, heat transport depends on the propagation of electrons and phonons, but it is also often analyzed using the thermal diffusion equation [109, 110]. For instance, dielectric multi-shell structures exhibit collective plasmon resonances across different metallic layers, resulting in a significant near-field enhancement of ~ $10^4$ at the core region, where light absorption is maximized [100, 108]. This confined nanoscale space creates a unique thermal environment under high pressure, providing new avenues for exploring novel physical or chemical processes.

### 3.2.4 *Interparticle Coupling*

Plasmon resonance is sensitive to the presence of nearby plasmonic structures, with the gap distance between nanostructures serving as a crucial parameter influencing the resonance [111–113]. When one plasmonic particle is in close proximity to another, their near-field interaction generates new plasmon-coupling resonances, significantly enhancing the electromagnetic field. A typical example of this phenomenon is observed in a particle dimer (Fig. 3.7a), where the dipolar plasmon mode splits into bonding and antibonding modes [80, 114, 115]. The bonding mode features two dipoles aligned along the interparticle axis, leading to strong dipole coupling with the far field. Conversely, the antibonding mode contains anti-aligned dipoles, resulting in a zero net dipole moment. Consequently, the optical response of the dimer is mainly governed by the lower-energy bonding mode.

For longitudinal excitation, the resonance band of the bonding mode undergoes a red shift as the gap distance decreases. This reduction in gap distance induces higher-order plasmon modes, which affect the resonance frequency, resulting in band splitting and broadening. The relative shift $\Delta\lambda/\lambda_0$ is exponentially related to the ratio of the gap to the diameter (*s/D*) as follows [7, 116]:

$$\frac{\Delta\lambda}{\lambda_0} = A\exp\left(-\frac{s}{\tau D}\right) \tag{3.1}$$

here $A$ represents an exponential factor, and $\tau$ is the decay constant. This type of decay is independent of the constituent materials, as well as the size and shape of particles. Figure 3.7b illustrates the spectral evolution of gold dimers with a diameter of 120 nm while varying the interparticle distance [117]. When the interparticle distance is greater than $3r$, the plasmon-coupling interaction is weak, resulting in a single-dipole mode across the spectrum. As the gap distance continues to decrease and approaches contact, the bonding mode initially undergoes a redshift. Subsequently, a higher-order quadrupolar resonance mode appears in the plasmon spectrum. However, the range over which plasmon resonance can broaden is constrained for a dimer.

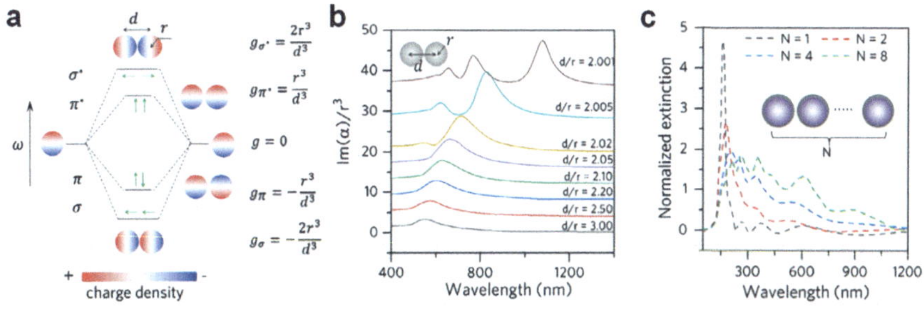


**Fig. 3.7** Interparticle coupling: **a** Energy hybridization within a plasmonic dimer. **b** Absorption spectra for varying interparticle distances. **c** Normalized extinction cross section ($C_{ext}/C_{geo}$) as a function of particle number. Adapted from Ref. [80]. Copyright 2021 Wiley–VCH

To cover a broad solar spectrum, plasmon coupling involving multiple interacting particles is achieved by assembling them into porous films with distributed sizes (Fig. 3.7c) [118]. Within these films, the plasmonic components are densely packed, causing their individual plasmon modes to overlap and hybridize. This leads to broadband light absorption [119]. The plasmon hybridization results in a noticeable redshift of the resonant band toward the visible-to-NIR regime. This method paves the way for innovative strategies to enhance photothermal conversion through structural engineering of nanoplasmon assemblies. Plasmonic thin films are also engineered for improved photothermal conversion by self-assembling particles onto substrates such as paper [120–124]. The particles are randomly distributed within the pores, creating a high density of hybridized modes that enable broad solar energy absorption. Furthermore, the porous structure minimizes back reflection of light. As a result, these films absorb approximately 99% of light across the full spectrum and demonstrate photothermal conversion efficiency exceeding 90%.

Both the electric field and plasmonic heating evidently depend on interparticle distance as well as the plasmon resonance [79, 80, 125]. The field enhancement factor can increase dramatically, up to $10^6$, as the interparticle distance decreases to atomic scales [126–128]. This suggests that significant field enhancement can be achieved by coupling particles with very small separations. However, the increase in temperature does not perfectly correlate with the ratio of light absorption powers. Instead, the temperature variation closely follows the profile of the dominant plasmon mode, exhibiting a distinct characteristic at the plasmon wavelength [129]. Consequently, the local temperature of a specific particle can be either lower or higher than adjacent particles, depending on the incident wavelength and specific interparticle distances [119]. For heating effects, both localized and uniform temperature distributions can be achieved by regulating the total number of nanoparticles, their interparticle distances, and the illumination conditions. The substrates and surrounding environment significantly influence the resulting temperature distribution, as they govern the rate and extent of heat transfer induced by localized surface plasmon resonance [79, 130]. Consequently, by carefully designing NP assemblies and implementing thermal management through substrate engineering, both localized and collective heating effects can be effectively harnessed for desired applications.

### *3.2.5 Active Coupling*

Actively tunable plasmonics critically modulate thermal, optical, electronic, and catalytic properties by the integration of plasmonic nanoparticles with stimuli-responsive components [7, 131, 132]. These reconfigurable materials employ a diverse array of active media including photochromic molecules, optically active J-aggregates, and quantum dots; thermoresponsive substances like vanadium oxide, gallium, and germanium antimony telluride; and electrically responsive materials such as liquid crystals, graphene, and elastic polymers [133–135]. Upon external stimuli such as light, heat, chemicals, or electricity, the elements are transied between states to optimize system energy. This dynamic adjustment in the coupling of plasmonic particles, mediated by the tunable surrounding medium, alters both plasmon resonance and the electromagnetic field.

The state transition mechanism is fascinating due to its implications for molecular dynamics, energy transport, and nanomechanics. A direct method to regulate plasmon response involves using materials with adaptable dielectric properties in response to light [136]. This regulation can result from photochemical reactions such as cis–trans isomerization, pericyclic reactions, intramolecular group transfers, hydrogen transfers, and dissociations [137]. Materials responsive to electrochemical stimuli facilitate reversible redox reactions, effectively tuning electronic structures and modulating their dielectric functions [138]. The dynamic manipulation of electro-optic films by electrical stimuli actively alters the refractive index, thereby adjusting the plasmon propagation constant and enabling precise control over light reflection [139, 140]. Chemical stimuli provide diverse methods for plasmon control, potentially through changes in ionic strength, pH, humidity, and specific DNA interactions [141, 142]. Thermally responsive materials exhibit continuous refractive index variations with temperature changes, enabling reversible and dynamic plasmon tuning [143, 144].

Thermo-optical materials stand out among thermoresponsive substances due to their rapid responses and low state transition thresholds [137]. The efficient photothermal conversion of plasmonic nanostructures presents promising avenues for light-directed manipulation of nanoparticle assemblies (Fig. 3.8), which depicts PNIPAM-coated gold particles [145]. Here, temperature around the particles decreases inversely with distance, contrasting with bulk heating, where plasmonic enhancement localizes thermal energy at the nanoscale for more effective energy conversion and accelerated thermal responses. Optical heating above a critical temperature expels water from core–shell particle coatings, triggering a switch in PNIPAM from hydrophilic to hydrophobic. This shift facilitates controlled aggregation of nanoparticles into mesoscale clusters, reversibly changing their color from red to dark blue (Fig. 3.8a). Accordingly, the plasmon resonance peak in the extinction spectra shifts between 525 and 645 nm, enabling reversible control (Fig. 3.8b).

Integrating plasmonics with specific polymers or molecules facilitates molecular conformational changes or reversible self-interactions due to external stimuli [7]. These plasmonic structures can be dispersed in solutions or embedded in solid matrices. By varying organic molecule lengths [146], the intergap distance between active components can be adjusted, influencing spectral response and thermoplasmonic heating through the shapes and sizes of the plasmonic structures. For instance, charged Au particles in solution self-assemble into linear structures [131, 147], displaying unique optical responses driven by the interplay of electrostatic repulsion and van der Waals attraction [148]. The influence of electrostatic forces varies with temperature and solution pH. Reversible assembly of gold nanorods, facilitated by pH adjustments and mediated by thiol-containing bifunctional molecules, has been demonstrated [149]. This leads to the reconfiguration of plasmon assemblies from an end-to-end alignment to a side-by-side arrangement. In the former case,

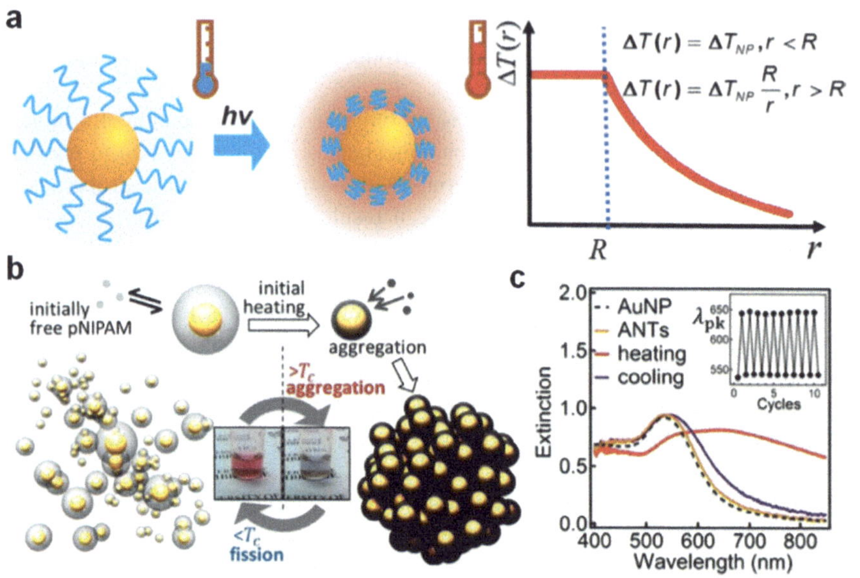


**Fig. 3.8** Plasmonics with adjustable dielectric environments: **a** temperature profiling of an Au particle under light irradiation, **b** reversible manipulation of thermoplasmonics via external illumination, and **c** extinction spectra demonstrating reversible heating and cooling cycles. Adapted from Ref. [145]. Copyright 2016 National Academy of Sciences

there is a red shift in the plasmon band featuring a longitudinal peak at 220 nm, whereas the latter alignment exhibits a 30 nm blue shift.

Responsive polymers often serve as a soft matrix to host plasmonic structures, acting as capping agents to inhibit particle aggregation. The entrapment of dense plasmonic structures within these matrices is responsible for their optical spectral responses [150]. Under external mechanical stimuli, dynamic couplings have been observed through alterations in the substrate's elastic properties. Stretching results in particle dispersion along the stretching axis and compaction perpendicular to it, leading to a plasmon band redshift from 510 to 520 nm with substrate elongation from 0 to 27%, thereby introducing polarization-dependent plasmonic properties (Fig. 3.9a) [151, 152]. This stretch-induced chain formation generates electromagnetic fields and thermal hot spots, raising the temperature from 26.5 °C at rest to 41.3 °C under 27% strain (Fig. 3.9b) [152]. These findings highlight the plasmo-mechanical system's dependence on light polarization, interparticle distance, and spatial particle arrangement.

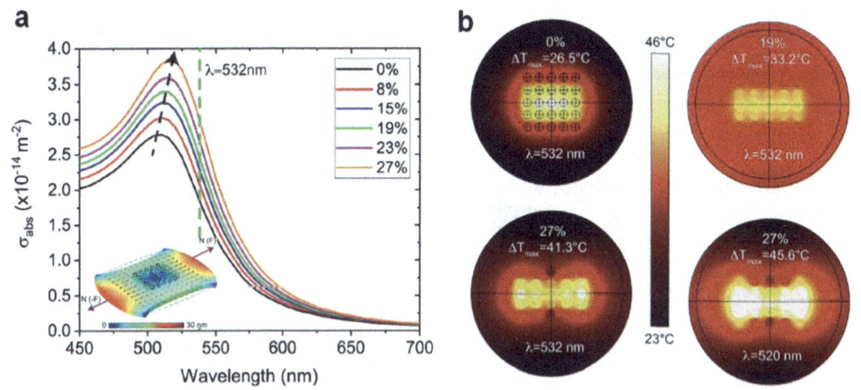


**Fig. 3.9** Thermoplasmonics featuring adjustable gap upon stretching: **a** absorption cross section curves, and **b** temperature mapping across varying degrees of stretch. Reproduced with permission from Ref. [152]. Copyright 2020 AIP

## 3.3 Material Fabrication

Investigating thermoplasmonic properties requires fabrication of nanoparticles with specific sizes and geometries. The production process adheres to either top-down (utilizing semiconductor industry microfabrication) or bottom-up (employing chemical synthesis) strategies [153]. Here, we offer a synopsis of frequently used methodologies, highlighting the universal challenges intrinsic to these diverse techniques.

### *3.3.1 Top-Down: Lithography*

Top-down techniques follow the strategy including template fabrication and metal deposition. A predefined pattern is first created on a flat substrate and then transferred onto a thin metal film using either additive or subtractive methods. The additive process involves depositing metal into cavities of a patterned template, which is later removed to reveal the underlying metal pattern. The subtractive process positions a template over a continuous metal film; unprotected metal areas are etched away. However, most technologies are limited to producing two-dimensional structures on flat substrates. Optical diffraction limits the processing resolution, which is unsuitable for constructing delicate plasmonic arrays [154]. Based on pattern generation, the techniques fall into four categories (Fig. 3.10): nanosphere (NSL) [155], nanoimprint (NIL) [156], holographic (HL),[157] and direct writing (DWL) [158, 159], Each technique possesses its own merits and demerits. It is crucial to select the appropriate method in accordance with the purpose of specific application.

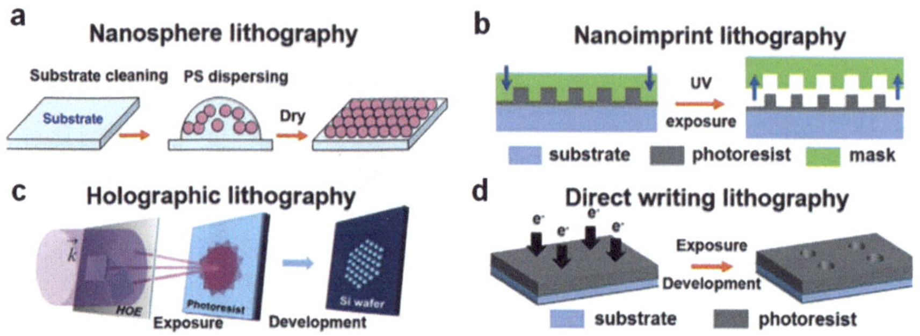


**Fig. 3.10** Template fabrication involves several techniques: **a** Nanosphere; **b** Nanoimprint; **c** Holographic; **d** Direct writing. Reproduced with permission from Ref. [158]. Copyright 2021 Wiley–VCH GmbH

NSL technique employs capillary forces among spherical colloids to generate closely packed structural patterns (Fig. 3.10a) [155], facilitated by methods such as Langmuir–Blodgett assembly, dip coating, and spin coating, yielding monolayer patterns that serve as masks for fabricating nanostructured plasmonic arrays [160]. Despite its effectiveness, achieving precise colloidal particle assembly presents a challenge, often leading to defects in the plasmonic arrays and affecting the uniformity of the morphology [161, 162]. Enhancements in uniformity and defect reduction can be achieved by integrating NSL with anodic aluminum oxide membranes and plasma etching techniques [163].

NIL method utilizes a mechanically prefabricated stamp to imprint a specific pattern onto a soft resist layer on a substrate, followed by UV light irradiation to harden the resist [156, 164]. This process effectively transfers the desired pattern into the resist, enabling the fabrication of plasmonic arrays (Fig. 3.10b). Demonstrating processing precision below 10 nm, NIL is economically advantageous as the reusable stamp can significantly reduce production costs [165]. However, the quality and resolution of the resultant structures depend critically on the characteristics of the prefabricated stamp.

HL, known as interference lithography, is a maskless technique that utilizes the interference patterns created by multiple laser beams to form specific designs (Fig. 3.10c) [166]. This approach allows the photoresist on a solid substrate to be simultaneously exposed over a large area to the interference pattern, resulting in structures with high uniformity. By adjusting factors such as the number and intensity of lasers, the angle of incidence, and the exposure duration, it is possible to modify the morphology and size of the individual units within the array [167, 168]. Thus, HL serves as an effective method for producing plasmonic nanostructure arrays. The achievable resolution of this technique is limited to the half-wavelength of the laser used, suggesting the use of deep UV lasers or immersion techniques for nanoscale processing resolution [169].

DWL involving focused ion beam (FIB), electron beam lithography (EBL), and laser printing (LP) uses precise techniques for pattern creation [170–172]. FIB and

EBL inscribe specific patterns onto photoresist using electron and focused ion beams (Fig. 3.10d); however, these point-scan techniques are low throughput and are thus primarily deployed for proof-of-concept demonstrations [173]. LP employs ultra-short laser pulses focused through an objective lens to create templates, capable of exposing photoresist as well as melting and reshaping materials like dielectrics or metals [172]. The LP efficiency can be improved by using a beam with micro-lens array, but its spatial resolution remains constrained by diffraction limits [174]. This limitation can be partially overcome by advanced techniques such as two-photon lithography or photothermal principles [175], which offer potential for fabricating structures down to sub-10 nm dimensions [176, 177].

Following template fabrication, metal deposition employs direct deposition or deposition and dissolution for forming metallic arrays (Fig. 3.11). Direct deposition precisely mirrors the template morphology onto a nanostructured array. This process can be further divided as shown in Fig. 3.11a. Using magnetron sputtering as an isotropic method, metals uniformly coat the structure, enveloping the base of the template and producing a plasmonic nanoarray marked by a continuous layer [178]. In contrast, anisotropic methods like electron beam or thermal evaporation or result in metal deposition that avoids the template's sidewalls, creating plasmonic strcutures with discontinuous metal layers, leaving parts of the template exposed (Fig. 3.11b) [179].

The deposition and dissolution is often utilized to replicate the template morphology (Fig. 3.11c), beginning with the deposition of a metal layer onto a substrate [180]. This is followed by sequential application of a sacrificial layer and a photoresist layer. The photoresist is then patterned to shape the template, which is etched into the sacrificial layer [181]. Subsequent anisotropic metal deposition precedes the removal of the sacrificial layer via a lift-off process, yielding a detailed metallic plasmonic array. This technique is important for the fabrication of plasmonic nanoarrays, enabling the incorporation of diverse materials such as semiconductors and two-dimensional substances that enhance the functionalities of the arrays [182,

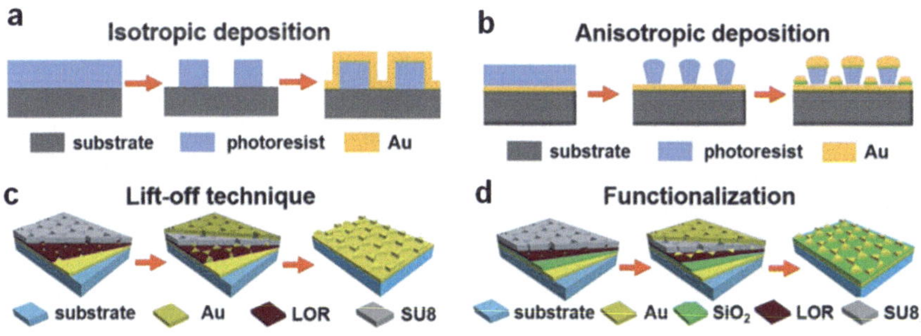


**Fig. 3.11** Metal deposition techniques: **a** uniform coating by isotropic deposition, **b** selectively deposits metal by anisotropic deposition; **c** pattern transfer via lift-off technique; **d** enhancing surface properties via functionalization. Adapted from Ref. [158]. Copyright 2021 Wiley–VCH GmbH

183]. For instance, a metal–dielectric–metal configuration might commence with the deposition of a dielectric layer, like $SiO_2$, onto a metal film, followed by layers added through further deposition and dissolution steps to form a structured sandwich-like assembly (Fig. 3.11d) [181].

### 3.3.2 Bottom-Up: Colloidal Synthesis

Considerable advancements are being made in bottom-up techniques due to the top-down approaches require large-scale equipment and cleanroom environments. Among these methods, colloidal synthesis is adept at producing nanoparticles with well-defined sizes and shapes that exhibit pronounced plasmonic resonances [184]. Moreover, this approach facilitates the creation of three-dimensional structures with precise nanoscale control [185, 186]. However, the particle shapes and assemblies achievable via colloidal synthesis remain relatively limited.

*Spherical particles*: Faraday devised a colloidal synthesis method whereby a metal salt is dissolved in a solvent, followed by the reduction of metal ions to their zero-valence state by using of a reducing agent [187]. This initiates the formation of metal nuclei within the solution, growing as more metal is reduced until the desired nanoparticle size is achieved. A key example includes the reaction between chloroauric acid and sodium citrate in water, where Au(III) ions are reduced by negative charged citrate ions [188]. These ions also attach to the particle surfaces, forming quasi-spherical shapes and moderating growth by capping the particles. This allows separate control over nucleation and growth through fine-tuning of reagent concentrations and reaction conditions, thus enabling precise particle size control and achieving a narrow size distribution (Fig. 3.12) [189]. Uniformity in particle size is crucial for plasmonic applications. The citrate's water solubility and negative charge contribute to stable suspension and prevent aggregation, although this charge-dependent stabilization is easily disrupted by variations in salt concentration, pH, temperature, and particle concentration, due to weak adsorption strength of citrate on the particle surfaces [131].

Conversely, synthesizing nanoparticles with organic capping ligands that form covalent bonds significantly enhances their stability [190]. A notable technique developed by Mathias Brust encompasses the reduction of chloroauric ions utilizing sodium borohydride within a toluene medium. This method incorporates dodecanethiol, or analogous long-chain alkanethiols as stabilizing capping agents. The synthesis transpires within an organic solvent and necessitates the translocation of $AuCl^{-4}$ ions from aqueous phase into the organic phase via a phase-transfer catalyst. The alkanethiol ligands serve as a steric barrier, preventing contact between gold surfaces of adjacent particles and thereby ensuring solution stability. This stabilization allows for improved monodispersity via post-synthesis processing, often utilizing size-selective precipitation to isolate monodisperse fractions from polydisperse samples. These particles can be segregated through centrifugation, and the

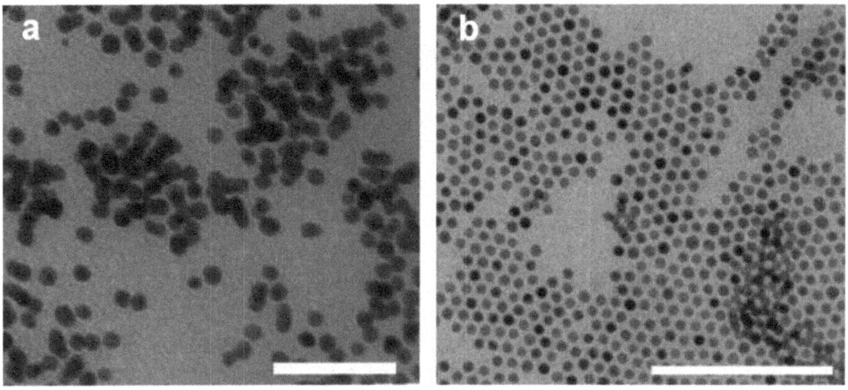


**Fig. 3.12** **a** TEM image of gold particles derived by citrate reduction in water; scale bar at 50 nm. **b** Gold particles capped with alkanethiol; scale bar at 100 nm. Reproduced with permission from Ref. [189] Copyright 2013 John Wiley & Sons

process may be repeated to achieve precise separations of several fractions of particles of varying sizes.

The most Brust derivatives produce small particles, ranging from 1 to 10 nm in size. To achieve monodisperse particles with controllable diameters, additional methods have been developed. The achievement of monodispersity needs the attenuation of coagulation, a process in which smaller particles amalgamate into larger aggregates, as well as Ostwald ripening, characterized by the migration of metal from smaller to larger particles. The use of capping agents that adhere to particle surfaces plays a crucial role in reducing the phenomena [191]. It often entails decoupling the rapid nucleation of seed particles from their controlled growth. In such systems, the particle concentration primarily depends on the initial number of nuclei, while individual particle size is controlled by the amount of reactant consumed until reaction completion. Hence, achieving size uniformity hinges on balancing kinetic processes with thermodynamic principles. For instance, the citrate technique is susceptible to thermodynamics promoting Ostwald ripening, potentially resulting in nonuniform particle sizes. However, monodispersity can be achieved through precise control over growth kinetics. Conversely, the Brust process targets specific particle sizes by maintaining equilibrium between the predefined concentrations of gold atoms and ligands. Furthermore, digestive ripening facilitates this equilibrium by redistributing material among particles of different sizes.

*Anisotropic particles*: Sharp-edged particles enhance the tunability of plasmonic resonance frequencies. Anisotropic shapes are in competition with surface tension, which favors minimizing surface area relative to volume. Silver and gold's face-centered cubic (fcc) lattice inherently impedes the formation of uniform, spherical single crystal particles. Instead, a balance between surface tension and this lattice type stabilizes symmetrical structures, notably the truncated octahedron, at zero temperature in a vacuum. Modifying surface energies through capping ligands can stabilize varying structures [185]. Additionally, twinning planes in nanoparticles offset the cubic symmetry, resulting in structures of lowered symmetry. Anisotropic structures generally emerge from growth processes conducted far from equilibrium. Effective synthesis of such structures often utilizes blocking groups to inhibit reactions at specific sites while facilitating them elsewhere. This approach involves capping agents that bind preferentially to specific crystal facets, promoting accelerated growth on alternate faces [192]. Often, particle shapes arise incidentally, with mechanisms being explained post hoc.

A prime example is the polyol synthesis of silver particles [15, 193]. In such method, ethylene glycol serves dual roles as both solvent and reducing agent for silver nitrate at elevated temperatures. Polyvinylpyrrolidone (PVP), employed as a capping agent, significantly influences the formation of nonspherical silver nanoparticles. PVP exhibits a stronger affinity for the {100} crystal facets of silver, stemming from differences in surface atom densities between the {100} and {111} facets. This differential binding facilitates preferential silver atom deposition on the {111} facets, promoting the diminution of these less stable surfaces, consequently forming nanocrystals predominantly bounded by {111} facets.

The synthesized structure is influenced by various synthetic parameters, with the seed structure formed immediately post-nucleation playing a critical role (Fig. 3.13). Single-crystal seeds generate platonic shapes like tetrahedra, octahedra, and cubes. Yet, multiply-twinned decahedral seeds are more common due to the low energy needed for integrating twinning planes within the silver lattice. These nuclei often form structures with five-fold symmetry by incorporating multiple twinning planes. Mackay icosahedra originate from the dense packing of spheres centered around a nucleus, serving as equilibrium seeds for the growth of particles with five-fold symmetry. As these particles grow, the icosahedral structure undergoes substantial strain and transitions into a decahedral configuration, distinguished by five twinning planes radiating from a central axis. These decahedra usually undergo morphological changes, elongating along their faces to form structures such as nanowires and nanorods.

Enhanced tuning of particle morphology can be achieved by separating the nucleation and growth stages. Initially, seed particles are synthesized independently and subsequently enlarged into larger, anisotropic structures through controlled reduction, often in the presence of PVP. Selective oxidation techniques can purify the initial solution by selectively etching multiply-twinned seeds, thus yielding single-crystal seeds. Under specific conditions, these seeds may exhibit multiple parallel twinning planes or stacking faults, diverging from the conventional radial twinning associated with five-fold symmetry. Conditions promoting slow growth are conducive to the formation of flat, plate-like nanoparticles, which may display hexagonal or triangular cross sections. Various colloidal growth methodologies have been refined to produce high yields of thin, triangular plates.

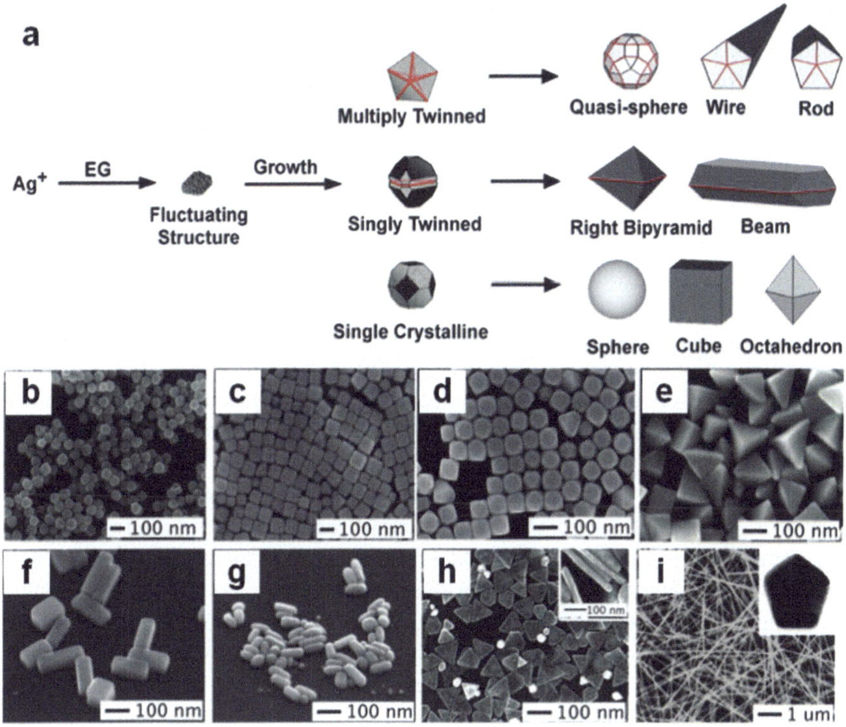


**Fig. 3.13** Synthesis of anisotropic particles. **a** Diagrams showing diverse particle shapes. **b–i** the crafted silver particles showcasing: **b** spherical, **c** cubic, **d** cuboids with truncated edges, **e** bipyramids with a right-angle configuration, **f** elongated bars, **g** ellipsoid shapes, **h** triangular flat forms, and **i** nanowires. Adapted from Ref. [15]. Copyright 2011 American Chemical Society

### *3.3.3 Self-assembly and Hybrid Approaches*

Chemical synthesis facilitates producing various 3D plasmonic particles. In contrast to lithographic methods, which can precisely organize particles into specified patterns, chemically synthesized particles tend to be dispersed randomly within a solution. Aligning these particles into specific configurations is crucial for many of their advanced applications [194, 195]. This section explores the methodologies that facilitate the organization of quantities of nanoparticles into predetermined geometrical configurations. We examine bottom-up strategies anchored in self-assembly principles, alongside hybrid approaches that integrate the merits of both top-down and bottom-up techniques.

One strategy for particle assembly uses Langmuir–Blodgett techniques, where nanoparticles suspended in a nonpolar organic solvent are spread on the water surface. As the solvent evaporates, hydrophobic molecules like alkanethiols on the particles form a sparse monolayer [196, 197]. Through meticulous compression of this layer, a dense nanoparticle monolayer develops, with adjustable interparticle spacing determined by the compression force. The Langmuir–Schaefer technique enables seamless transfer of this monolayer to various substrates, whereas the Langmuir–Blodgett method typically transfers nanoparticle lines due to the stick–slip movement of the liquid meniscus when vertically lifting the substrate from the trough [198]. Metal nanoparticles, because of their size, diffuse slowly across the surface, making highly ordered and compressed configurations challenging. Such films exhibit considerable short-range order but limited long-range order, usually spanning only 10–100 particles. Achieving significant order requires highly monodisperse colloidal samples, and without stabilization, these films disintegrate once surface pressure reduces. Stability can be enhanced by integrating binding molecules that connect adjacent nanoparticles. For quasi-spherical particles, a hexagonal lattice formation is typical, whereas cubic silver particles form square lattices, and octahedra create hexagonal arrangements [197]. Extremely anisotropic structures, like nanorods and nanowires, display liquid–crystal arrangements with smectic organization in short aspect ratio rods and nematic-like alignment in nanowires and rods with high aspect ratios.

Hybrid techniques merge the strengths of both top-down and bottom-up approaches. For instance, one method leverages self-organization to establish a template that subsequently directs metal deposition. This process mirrors lithographic patterning combined with lift-off, yet diverges as the shaping of the metal deposition and lift-off template is achieved through a bottom-up strategy rather than a top-down technique [199, 200]. An illustrative case is the crafting of metal nanoshells on colloidal silica particles, where the silica serves as a three-dimensional scaffold for the electroless deposition of metal [201]. Typically, templated deposition mechanisms entail the creation of a hole-filled array, later infilled with metal. A self-organized template is formed by a two-dimensional colloidal crystal of polymer or glass spheres, achieved by controlled evaporation of a silica or polymer colloid solution onto a substrate, yielding an orderly particle matrix. Metallic structures are finally created by depositing metal over the surface and then removing the spheres. While in selected dissolvable solvents, both polymer and glass spheres can be removed through ultrasound treatment of the sample.

Another example of a self-organized template involves hole arrays in anodic aluminum oxide (AAO) films [202]. Anodization serves as the reverse process to electroplating. When aluminum exposed to an acidic solution, the acid gradually dissolves the oxide layer. To fabricate an aluminum oxide membrane with uniformly pores arranged in a hexagonal pattern, it is crucial to balance the oxidation and etching rates. The pore size can be adjusted by altering the applied voltage, which ranges in 5–200 nm, and the membrane thickness varies from 10 to 100 μm [203, 204]. Filling these nanoporous membranes with metal converts them into a matrix of nanowires encased within the alumina layer. This conversion typically involves initially depositing a metal layer via evaporation or sputtering onto one side of the membrane, which then acts as an electrode for electroplating the metal into the pores. Controlling the process duration allows for tuning the nanowire length with aspect ratios from one to over 10,000. A variety of materials including different metals and nonmetals such as semiconductors can be deposited in this way. Additionally,

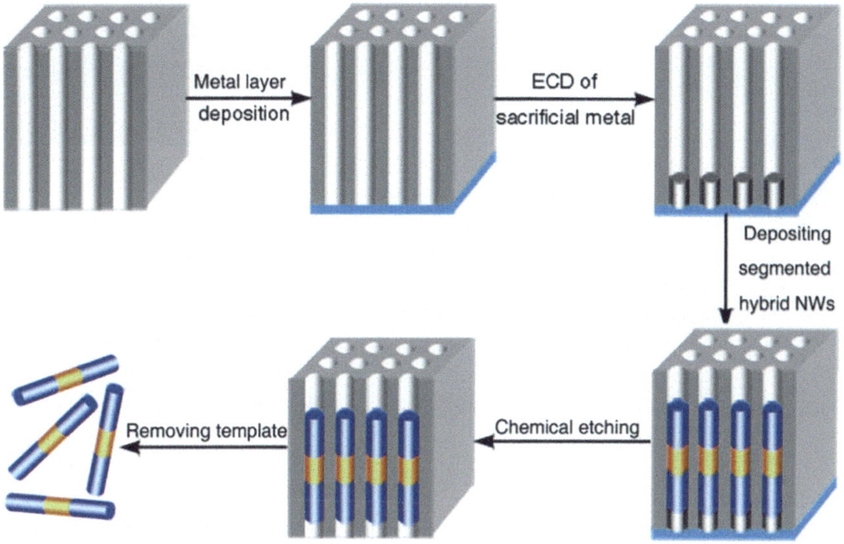


**Fig. 3.14** General procedure for electrochemical deposition (ECD) of segmented hybrid nanowires within AAO membranes. Adapted from Ref. [202]. Copyright 2018 Elsevier

it is feasible to switch materials mid-deposition to create nanowire heterostructures (Fig. 3.14) [202]. Alternatively, electroless deposition provides a means to form hollow metal tubes instead of solid wires, as the deposition initiates from the pore walls inward. The alumina template can then be selectively removed through chemical etching.

Enhanced flexibility is also achieved through a combine bottom-up approach used to synthesize colloidal particles, with top-down techniques for patterning substrates [205]. Lithographic methods enable the creation of grooves on a substrate, directing the assembly of colloidal particles via controlled evaporation (Fig. 3.15) [206]. Positioning the substrate at an angle within a colloidal solution leads to gradual evaporation, prompting a systematic descent of the solid–liquid contact line. Alternatively, using microfluidic flow cell or spin coater facilitates manipulation of the contact line across the substrate or even distribution of the solution across the patterned area. As the contact line traverses the surface, it anchors at topographic features, consequently enhancing local concentration and promoting deposition at these sites [207]. The anchorage duration is influenced by the wetting level between the solution and the substrate, indicated by the liquid's contact angle with the substrate. In aqueous solutions, adjusting this contact angle is possible through surfactant inclusion or substrate treatment via chemical functionalization, ozone exposure, or oxygen-plasma etching. Optimal contact angles, alongside proper particle concentration and precise feature heights, ensure selective particle deposition within the patterned indentations on the substrate.

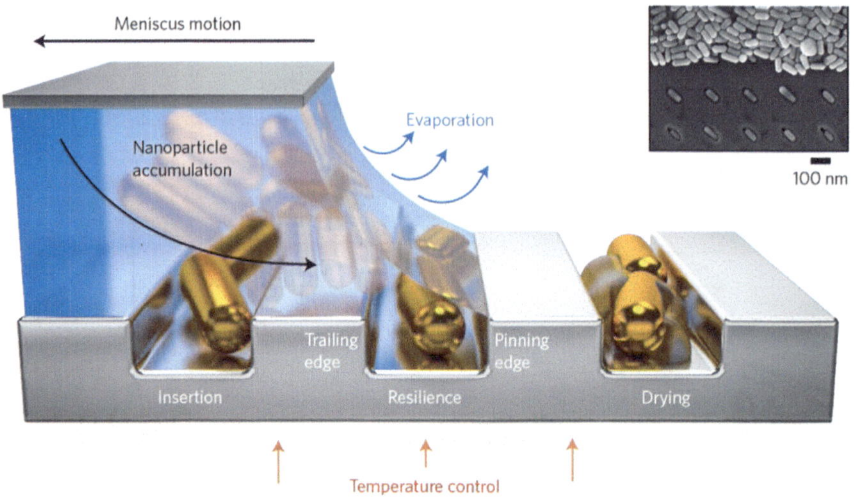


**Fig. 3.15** Capillary assembly of nanoparticles into topographical traps: (1) insertion of particles into the trap, (2) retention within the trap against removal by capillary doctor blading, and (3) maintenance of particle stability against displacement and reorientation during final solvent drying. Adapted from Ref. [206]. Copyright 2017 Nature Publishing Group

# References


1. G. Baffou, *Thermoplasmonics Heating Metal Nanoparticles Using Light* (Cambridge University Press, 2018). ISBN 978-971-108-41832-41834
2. S.A. Maier, *Plasmonics Fundamentals and Applications @2007* (Springer Science+Business Media LLC, 2007)
3. R.J. Mendelsberg, G. Garcia, D.J. Milliron, Extracting reliable electronic properties from transmission spectra of indium tin oxide thin films and nanocrystal films by careful application of the Drude theory. J. Appl. Phys. **111** (2012)
4. A. Boltasseva, H.A. Atwater, Low-loss plasmonic metamaterials. Science **331**, 290–291 (2011)
5. T. Gong, J.N. Munday, Materials for hot carrier plasmonics. Opt. Mater. Expr. **5**, 2501 (2015)
6. S. Furumi, H. Fudouzi, T. Sawada, Self-organized colloidal crystals for photonics and laser applications. Laser Photon. Rev. **4**, 205–220 (2010)
7. N. Jiang, X. Zhuo, J. Wang, Active plasmonics: principles, structures, and applications. Chem. Rev. **118**, 3054–3099 (2018)
8. A. Agrawal, R.W. Johns, D.J. Milliron, Control of localized surface plasmon resonances in metal oxide nanocrystals. Annu. Rev. Mater. Res. **47**, 1–31 (2017)
9. P.R. West, S. Ishii, G.V. Naik, N.K. Emani, V.M. Shalaev, A. Boltasseva, Searching for better plasmonic materials. Laser Photonics Rev. **4**, 795–808 (2010)
10. G.V. Naik, V.M. Shalaev, A. Boltasseva, Alternative plasmonic materials: beyond gold and silver. Adv. Mater. **25**, 3264–3294 (2013)
11. M.I. Stockman, K. Kneipp, S.I. Bozhevolnyi, S. Saha, A. Dutta, J. Ndukaife, N. Kinsey, H. Reddy, U. Guler, V.M. Shalaev, A. Boltasseva, B. Gholipour, H.N.S. Krishnamoorthy, K.F. MacDonald, C. Soci, N.I. Zheludev, V. Savinov, R. Singh, P. Gross, C. Lienau, M. Vadai, M.L. Solomon, D.R. Barton, M. Lawrence, J.A. Dionne, S.V. Boriskina, R. Esteban, J. Aizpurua, X. Zhang, S. Yang, D.Q. Wang, W.J. Wang, T.W. Odom, N. Accanto, P.M. de Roque, I.M. Hancu, L. Piatkowski, N.F. van Hulst, M.F. Kling, Roadmap on plasmonics. J. Optics-Uk **20** (2018)

12. A. Lalisse, G. Tessier, J. Plain, G. Baffou, Quantifying the efficiency of plasmonic materials for near-field enhancement and photothermal conversion. J. Phys. Chem. C **119**, 25518–25528 (2015)
13. H. Kang, J.T. Buchman, R.S. Rodriguez, H.L. Ring, J. He, K.C. Bantz, C.L. Haynes, Stabilization of silver and gold nanoparticles: preservation and improvement of plasmonic functionalities. Chem. Rev. **119**, 664–699 (2019)
14. H. Chen, L. Shao, Q. Li, J. Wang, Gold nanorods and their plasmonic properties. Chem. Soc. Rev. **42**, 2679–2724 (2013)
15. M. Rycenga, C.M. Cobley, J. Zeng, W. Li, C.H. Moran, Q. Zhang, D. Qin, Y. Xia, Controlling the synthesis and assembly of silver nanostructures for plasmonic applications. Chem. Rev. **111**, 3669–3712 (2011)
16. I. Zoric, M. Zach, B. Kasemo, C. Langhammer, Gold, platinum, and aluminum nanodisk plasmons: material independence, subradiance, and damping mechanisms. ACS Nano **5**, 2535–2546 (2011)
17. A. Marimuthu, J. Zhang, S. Linic, Tuning selectivity in propylene epoxidation by plasmon mediated photo-switching of Cu oxidation state. Science **339**, 1590–1593 (2013)
18. M.W. Knight, N.S. King, L. Liu, H.O. Everitt, P. Nordlander, N.J. Halas, Aluminum for plasmonics. ACS Nano **8**, 834–840 (2014)
19. B.D. Clark, C.R. Jacobson, M. Lou, J. Yang, L. Zhou, S. Gottheim, C.J. DeSantis, P. Nordlander, N.J. Halas, Aluminum nanorods. Nano Lett. **18**, 1234–1240 (2018)
20. A. Sobhani, A. Manjavacas, Y. Cao, M.J. McClain, F.J. Garcia de Abajo, P. Nordlander, N.J. Halas, Pronounced linewidth narrowing of an aluminum nanoparticle plasmon resonance by interaction with an aluminum metallic film. Nano Lett. **15**, 6946–6951 (2015)
21. Y. Zhu, L. Cui, M. Abbasi, D. Natelson, Tuning light emission crossovers in atomic-scale aluminum plasmonic tunnel junctions. Nano Lett. **22**, 8068–8075 (2022)
22. A.B. Urcan Guler, V.M. Shalaev, Refractory plasmonics. Science **344**, 263–264 (2014)
23. Z. Huang, C. Cao, Q. Wang, H. Zhang, C.E. Owens, A.J. Hart, K. Cui, Multiscale plasmonic refractory nanocomposites for high-temperature solar photothermal conversion. Nano Lett. **22**, 8526–8533 (2022)
24. X. Huang, S. Tang, X. Mu, Y. Dai, G. Chen, Z. Zhou, F. Ruan, Z. Yang, N. Zheng, Freestanding palladium nanosheets with plasmonic and catalytic properties. Nat. Nanotechnol. **6**, 28–32 (2011)
25. H. Sheng, J. Wang, J. Huang, Z. Li, G. Ren, L. Zhang, L. Yu, M. Zhao, X. Li, G. Li, N. Wang, C. Shen, G. Lu, Strong synergy between gold nanoparticles and cobalt porphyrin induces highly efficient photocatalytic hydrogen evolution. Nat. Commun. **14**, 1528 (2023)
26. S.S.V. Kailash, Quantifying the optical and thermoplasmonic properties of some bimetallic alloy nanospheres. J. Quant. Spectrosc. Radiat. Transfer **309**, 108707 (2023)
27. P. Han, P. Jin, X. Li, Y. Xu, K. Li, S. Wang, Z. Nie, Photoactivation of ambient oxygen via plasmon-coupled valence-band hybridization in AgPd nanoalloy for reaction pathway alteration. Appl. Catal. B **298**, 120598 (2021)
28. M. Rebello Sousa Dias, M.S. Leite, Alloying: a platform for metallic materials with on-demand optical response. Accounts Chem. Res. **52**, 2881–2891 (2019)
29. M. Cao, Q. Liu, M. Chen, L. Chen, D. Yang, H. Hu, L. He, G. Zhang, Q. Zhang, Fully alloying AuAg nanorods in a photothermal nano-oven: superior plasmonic property and enhanced chemical stability. ACS Omega **3**, 18623–18629 (2018)
30. N.E. Motl, A.F. Smith, C.J. DeSantis, S.E. Skrabalak, Engineering plasmonic metal colloids through composition and structural design. Chem. Soc. Rev. **43**, 3823–3834 (2014)
31. G. Baffou, *Thermoplasmonics Heating Metal Nanoparticles Using Light* (Cambridge University Press, 2017).
32. U. Guler, A. Boltasseva, V.M. Shalaev, Refractory plasmonics. Science **344**, 263–264 (2014)
33. C. Ma, J. Yan, Y. Huang, C. Wang, G. Yang, The optical duality of tellurium nanoparticles for broadband solar energy harvesting and efficient photothermal conversion. Sci. Adv. **4**, eaas9894 (2018)

34. W. Li, U. Guler, N. Kinsey, G.V. Naik, A. Boltasseva, J. Guan, V.M. Shalaev, A.V. Kildishev, Refractory plasmonics with titanium nitride: broadband metamaterial absorber. Adv. Mater. **26**, 7959–7965 (2014)
35. S.H.C. Askes, E.C. Garnett, Ultrafast thermal imprinting of plasmonic hotspots. Adv. Mater. **33**, e2105192 (2021)
36. S. Rej, L. Mascaretti, E.Y. Santiago, O. Tomanec, Š Kment, Z. Wang, R. Zbořil, P. Fornasiero, A.O. Govorov, A. Naldoni, Determining plasmonic hot electrons and photothermal effects during $H_2$ evolution with TiN-Pt nanohybrids. ACS Catal. **10**, 5261–5271 (2020)
37. A.N. Sabzeghabae, C. Berrospe-Rodriguez, L. Mangolini, G. Aguilar, Laser-induced cavitation in plasmonic nanoparticle solutions: a comparative study between gold and titanium nitride. J. Biomed. Mater. Res., Part A **109**, 2483–2492 (2021)
38. T. Krekeler, S.S. Rout, G.V. Krishnamurthy, M. Störmer, M. Arya, A. Ganguly, D.S. Sutherland, S.I. Bozhevolnyi, M. Ritter, K. Pedersen, A.Y. Petrov, M. Eich, M. Chirumamilla, Unprecedented thermal stability of plasmonic titanium nitride films up to 1400 °C. Adv. Opt. Mater. **9** (2021)
39. U. Guler, V.M. Shalaev, A. Boltasseva, Nanoparticle plasmonics: going practical with transition metal nitrides. Mater. Today **18**, 227–237 (2015)
40. T. Krekeler, S.S. Rout, G.V. Krishnamurthy, M. Störmer, M. Arya, A. Ganguly, D.S. Sutherland, S.I. Bozhevolnyi, M. Ritter, K. Pedersen, A.Y. Petrov, M. Eich, M. Chirumamilla, Unprecedented thermal stability of plasmonic titanium nitride films up to 1400 °C. Adv. Opt. Mater. 2100323 (2021)
41. S. Ishii, R.P. Sugavaneshwar, T. Nagao, Titanium nitride nanoparticles as plasmonic solar heat transducers. J. Phys. Chem. C **120**, 2343–2348 (2016)
42. Y. Guo, Z. Xu, A.G. Curto, Y.J. Zeng, D. Van Thourhout, Plasmonic semiconductors: materials, tunability and applications. Prog. Mater. Sci. **138**, 101158 (2023)
43. M. Gao, L. Zhu, C.K. Peh, G.W. Ho, Solar absorber material and system designs for photothermal water vaporization towards clean water and energy production. Energy Environ. Sci. **12**, 841–864 (2019)
44. J.T. Li, N.Q. Wu, Semiconductor-based photocatalysts and photoelectrochemical cells for solar fuel generation: a review, Catal. Sci Technol **5**, 1360–1384 (2015)
45. X. Li, D. Wang, Y. Zhang, L. Liu, W. Wang, Surface-ligand protected reduction on plasmonic tuning of one-dimensional $MoO_{3-x}$ nanobelts for solar steam generation. Nano Res. **13**, 3025–3032 (2020)
46. J. Wang, Y. Li, L. Deng, N. Wei, Y. Weng, S. Dong, D. Qi, J. Qiu, X. Chen, T. Wu, High-performance photothermal conversion of narrow-bandgap $Ti_2O_3$ nanoparticles. Adv. Mater. **29** (2017)
47. J.M. Luther, P.K. Jain, T. Ewers, A.P. Alivisatos, Localized surface plasmon resonances arising from free carriers in doped quantum dots. Nat. Mater. **10**, 361–366 (2011)
48. C.M. Hessel, V.P. Pattani, M. Rasch, M.G. Panthani, B. Koo, J.W. Tunnell, B.A. Korgel, Copper selenide nanocrystals for photothermal therapy. Nano Lett. **11**, 2560–2566 (2011)
49. K. Manthiram, A.P. Alivisatos, Tunable localized surface plasmon resonances in tungsten oxide nanocrystals. J. Am. Chem. Soc. **134**, 3995–3998 (2012)
50. W. Xu, H. Liu, D. Zhou, X. Chen, N. Ding, H. Song, H. Ågren, Localized surface plasmon resonances in self-doped copper chalcogenide binary nanocrystals and their emerging applications. Nano Today **33**, 100892 (2020)
51. K. Hantanasirisakul, Y. Gogotsi, Electronic and optical properties of 2D transition metal carbides and nitrides (MXenes). Adv. Mater. **30** (2018)
52. G.R. Bhimanapati, Z. Lin, V. Meunier, Y. Jung, J. Cha, S. Das, D. Xiao, Y. Son, M.S. Strano, V.R. Cooper, L.B. Liang, S.G. Louie, E. Ringe, W. Zhou, S.S. Kim, R.R. Naik, B.G. Sumpter, H. Terrones, F.N. Xia, Y.L. Wang, J. Zhu, D. Akinwande, N. Alem, J.A. Schuller, R.E. Schaak, M. Terrones, J.A. Robinson, Recent advances in two-dimensional materials beyond graphene. ACS Nano **9**, 11509–11539 (2015)
53. R. Yu, Q. Guo, F. Xia, F.J. Garcia de Abajo, Photothermal engineering of graphene plasmons. Phys. Rev. Lett. **121**, 057404 (2018)

54. Z. Fei, A.S. Rodin, G.O. Andreev, W. Bao, A.S. McLeod, M. Wagner, L.M. Zhang, Z. Zhao, M. Thiemens, G. Dominguez, M.M. Fogler, A.H. Castro Neto, C.N. Lau, F. Keilmann, D.N. Basov, Gate-tuning of graphene plasmons revealed by infrared nano-imaging. Nature **487**, 82–85 (2012)
55. X. Wen, W. Wang, X. Zhang, H. Chen, S. Jia, Y. Gong, W. Chen, Y. Wang, H. Zhu, J. Zheng, P.M. Ajayan, J. Lou, Pathways of exciton triggered hot-carrier injection at plasmonic metal-transition metal dichalcogenide interface. Adv. Opt. Mater. **10** (2022)
56. D. Yang, G. Yang, P. Yang, R. Lv, S. Gai, C. Li, F. He, J. Lin, Assembly of Au plasmonic photothermal agent and iron oxide nanoparticles on ultrathin black phosphorus for targeted photothermal and photodynamic cancer therapy. Adv. Funct. Mater. **27** (2017)
57. J. Shao, H. Xie, H. Huang, Z. Li, Z. Sun, Y. Xu, Q. Xiao, X.F. Yu, Y. Zhao, H. Zhang, H. Wang, P.K. Chu, Biodegradable black phosphorus-based nanospheres for in vivo photothermal cancer therapy. Nat. Commun. **7**, 12967 (2016)
58. D. Alcaraz Iranzo, S. Nanot, E.J.C. Dias, I. Epstein, C. Peng, D.K. Efetov, M.B. Lundeberg, R. Parret, J. Osmond, J.Y. Hong, J. Kong, D.R. Englund, N.M.R. Peres, F.H.L. Koppens, Probing the ultimate plasmon confinement limits with a van der Waals heterostructure. Science **360**, 291–295 (2018)
59. T. Low, A. Chaves, J.D. Caldwell, A. Kumar, N.X. Fang, P. Avouris, T.F. Heinz, F. Guinea, L. Martin-Moreno, F. Koppens, Polaritons in layered two-dimensional materials. Nat. Mater. **16**, 182–194 (2017)
60. Y. Wang, J.Z. Ou, A.F. Chrimes, B.J. Carey, T. Daeneke, M.M. Alsaif, M. Mortazavi, S. Zhuiykov, N. Medhekar, M. Bhaskaran, J.R. Friend, M.S. Strano, K. Kalantar-Zadeh, Plasmon resonances of highly doped two-dimensional $MoS_2$. Nano Lett. **15**, 883–890 (2015)
61. L. Zhang, J. Dong, F. Ding, Strategies, status, and challenges in wafer scale single crystalline two-dimensional materials synthesis. Chem. Rev. **121**, 6321–6372 (2021)
62. J. Pang, R.G. Mendes, A. Bachmatiuk, L. Zhao, H.Q. Ta, T. Gemming, H. Liu, Z. Liu, M.H. Rummeli, Applications of 2D MXenes in energy conversion and storage systems. Chem. Soc. Rev. **48**, 72–133 (2019)
63. A. VahidMohammadi, J. Rosen, Y. Gogotsi, The world of two-dimensional carbides and nitrides (MXenes). Science **372** (2021)
64. X. Li, Z. Huang, C.E. Shuck, G. Liang, Y. Gogotsi, C. Zhi, MXene chemistry, electrochemistry and energy storage applications, Nature reviews. Chemistry **6**, 389–404 (2022)
65. X. Guo, N. Li, C. Wu, X. Dai, R. Qi, T. Qiao, T. Su, D. Lei, N. Liu, J. Du, E. Wang, X. Yang, P. Gao, Q. Dai, Studying plasmon dispersion of MXene for enhanced electromagnetic absorption. Adv. Mater. **34**, e2201120 (2022)
66. N. Zhang, J. Zhang, X. Zhu, S. Yuan, D. Wang, H. Xu, Z. Wang, Synergistic effect of $Ti_3C_2T_x$ MXene nanosheets and tannic acid-$Fe^{3+}$ network in constructing high-performance hydrogel composite membrane for photothermal membrane distillation. Nano Lett. **24**, 724–732 (2024)
67. R. Li, L. Zhang, L. Shi, P. Wang, MXene $Ti_3C_2$: an effective 2D light-to-heat conversion material. ACS Nano **11**, 3752–3759 (2017)
68. Q. Wu, L. Tan, X. Liu, Z. Li, Y. Zhang, Y. Zheng, Y. Liang, Z. Cui, S. Zhu, S. Wu, The enhanced near-infrared photocatalytic and photothermal effects of MXene-based heterojunction for rapid bacteria-killing. Appl. Catal. B **297**, 120500 (2021)
69. Z. Hao, Y. Li, X. Liu, T. Jiang, Y. He, X. Zhang, C. Cong, D. Wang, Z. Liu, D. Gao, Enhancing biocatalysis of a MXene-based biomimetic plasmonic assembly for targeted cancer treatments in NIR-II biowindow. Chem. Eng. J. **425**, 130639 (2021)
70. A. Manjavacas, J.G. Liu, V. Kulkarni, P. Nordlander, Plasmon-induced hot carriers in metallic nanoparticles. ACS Nano **8**, 7630–7638 (2014)
71. G.S. Javier González-Colsa, J.M. Saiz, D. Ortiz, F. González, F. Bresme, F. Moreno, P. Albella, Gold nanodoughnut as an outstanding nanoheater for photothermal applications, arXiv:2106. 01059 (2021)
72. J. Qiu, M. Xie, T. Wu, D. Qin, Y. Xia, Gold nanocages for effective photothermal conversion and related applications. Chem. Sci. **11**, 12955–12973 (2020)

73. K. Bae, G. Kang, S.K. Cho, W. Park, K. Kim, W.J. Padilla, Flexible thin-film black gold membranes with ultrabroadband plasmonic nanofocusing for efficient solar vapour generation. Nat. Commun. **6**, 10103 (2015)
74. S.V. Boriskina, T.A. Cooper, L. Zeng, G. Ni, J.K. Tong, Y. Tsurimaki, Y. Huang, L. Meroueh, G. Mahan, G. Chen, Losses in plasmonics: from mitigating energy dissipation to embracing loss-enabled functionalities. Adv. Opt. Photon. **9**, 775 (2017)
75. C. Mouhot, C. Villani, On Landau damping. Acta Math. **207**, 29–201 (2011)
76. S. Peng, J.M. McMahon, G.C. Schatz, S.K. Gray, Y. Sun, Reversing the size-dependence of surface plasmon resonances. Proc. Natl. Acad. Sci. **107**, 14530–14534 (2010)
77. O. Hess, J.B. Pendry, S.A. Maier, R.F. Oulton, J.M. Hamm, K.L. Tsakmakidis, Active nanoplasmonic metamaterials. Nat. Mater. **11**, 573–584 (2012)
78. S. Link, M.A. El-Sayed, Spectral properties and relaxation dynamics of surface plasmon electronic oscillations in gold and silver nanodots and nanorods. J. Phys. Chem. B **103**, 8410–8426 (1999)
79. L. Jauffred, A. Samadi, H. Klingberg, P.M. Bendix, L.B. Oddershede, Plasmonic heating of nanostructures. Chem. Rev. **119**, 8087–8130 (2019)
80. J. Chen, Z. Ye, F. Yang, Y. Yin, Plasmonic nanostructures for photothermal conversion. Small Sci. **1**, 2000055 (2021)
81. V. Amendola, R. Pilot, M. Frasconi, O.M. Maragò, M.A. Iatì, Surface plasmon resonance in gold nanoparticles: a review. J. Phys. Condens. Matter **29**, 203002 (2017)
82. B. Yang, C. Li, Z. Wang, Q. Dai, Thermoplasmonics in solar energy conversion: materials. Nanostruct. Des. Appl. Adv. Mater. **34**, e2107351 (2022)
83. Y. Seol, A.E. Carpenter, T.T. Perkins, Gold nanoparticles: enhanced optical trapping and sensitivity coupled with significant heating. Opt. Lett. **31**, 2429 (2006)
84. E.R. Encina, E.A. Coronado, Size optimization of iron Oxide@Noble metal core-shell nanohybrids for photothermal applications. J. Phys. Chem. C **120**, 5630–5639 (2016)
85. H. Zhang, T. Zhu, M. Li, Quantitative analysis of the shape effect of thermoplasmonics in gold nanostructures. J. Phys. Chem. Lett. **14**, 3853–3860 (2023)
86. E.Y. Santiago, L.V. Besteiro, X.T. Kong, M.A. Correa-Duarte, Z. Wang, A.O. Govorov, Efficiency of hot-electron generation in plasmonic nanocrystals with complex shapes: surface-induced scattering, hot spots, and interband transitions. ACS Photon. **7**, 2807–2824 (2020)
87. G.A. Vinnacombe-Willson, N. Chiang, L. Scarabelli, Y. Hu, L.K. Heidenreich, X. Li, Y. Gong, D.T. Inouye, T.S. Fisher, P.S. Weiss, S.J. Jonas, In situ shape control of thermoplasmonic gold nanostars on oxide substrates for hyperthermia-mediated cell detachment. ACS Cent. Sci. **6**, 2105–2116 (2020)
88. H. Yuan, A.M. Fales, T. Vo-Dinh, TAT peptide-functionalized gold nanostars: enhanced intracellular delivery and efficient NIR photothermal therapy using ultralow irradiance. J. Am. Chem. Soc. **134**, 11358–11361 (2012)
89. H. Liu, T.E. Gage, P. Singh, A. Jaiswal, R.D. Schaller, J. Tang, S.T. Park, S.K. Gray, I. Arslan, Visualization of plasmonic couplings using ultrafast electron microscopy. Nano Lett. **21**, 5842–5849 (2021)
90. M.B. Ross, M.G. Blaber, G.C. Schatz, Using nanoscale and mesoscale anisotropy to engineer the optical response of three-dimensional plasmonic metamaterials. Nat. Commun. **5**, 4090 (2014)
91. G. Baffou, R. Quidant, C. Girard, Heat generation in plasmonic nanostructures: influence of morphology. Appl. Phys. Lett. **94** (2009)
92. E. Ringe, M.R. Langille, K. Sohn, J. Zhang, J. Huang, C.A. Mirkin, R.P. Van Duyne, L.D. Marks, Plasmon length: a universal parameter to describe size effects in gold nanoparticles. J. Phys. Chem. Lett. **3**, 1479–1483 (2012)
93. B.J. Wiley, S.H. Im, Z.Y. Li, J. McLellan, A. Siekkinen, Y. Xia, Maneuvering the surface plasmon resonance of silver nanostructures through shape-controlled synthesis. J. Phys. Chem. B **110**, 15666–15675 (2006)

94. O. Cardozo, M. Habib, W. Jiang, R.E. de Araujo, S. Farooq, Superior performance of hollow plasmonic cubic structures for solar energy harvesting, conversion, and storage systems. Plasmonics (2024)
95. B. Gao, G. Arya, A.R. Tao, Self-orienting nanocubes for the assembly of plasmonic nanojunctions. Nat. Nanotechnol. **7**, 433–437 (2012)
96. M. Ha, S.H. Nam, K. Sim, S.E. Chong, J. Kim, Y. Kim, Y. Lee, J.M. Nam, Highly efficient photothermal therapy with cell-penetrating peptide-modified bumpy au triangular nanoprisms using low laser power and low probe dose. Nano Lett. **21**, 731–739 (2021)
97. G. Baffou, C. Girard, R. Quidant, Mapping heat origin in plasmonic structures. Phys. Rev. Lett. **104**, 136805 (2010)
98. J. Reguera, J. Langer, D. Jimenez de Aberasturi, L.M. Liz-Marzan, Anisotropic metal nanoparticles for surface enhanced Raman scattering. Chem. Soc. Rev. **46**, 3866–3885 (2017)
99. J. Pettine, P. Choo, F. Medeghini, T.W. Odom, D.J. Nesbitt, Plasmonic nanostar photocathodes for optically-controlled directional currents. Nat. Commun. **11** (2020)
100. L. Meng, R. Yu, M. Qiu, F.J. Garcia de Abajo, Plasmonic nano-oven by concatenation of multishell photothermal enhancement. ACS Nano **11**, 7915–7924 (2017)
101. M.A. Mahmoud, M. Chamanzar, A. Adibi, M.A. El-Sayed, Effect of the dielectric constant of the surrounding medium and the substrate on the surface plasmon resonance spectrum and sensitivity factors of highly symmetric systems: silver nanocubes. J. Am. Chem. Soc. **134**, 6434–6442 (2012)
102. L. Wang, M. Hasanzadeh Kafshgari, M. Meunier, Optical properties and applications of plasmonic-metal nanoparticles. Adv. Funct. Mater. **30**, 2005400 (2020)
103. M. ElKabbash, A. Sousa-Castillo, Q. Nguyen, R. Mariño-Fernández, N. Hoffman, M.A. Correa-Duarte, G. Strangi, Tunable black gold: controlling the near-field coupling of immobilized au nanoparticles embedded in mesoporous silica capsules. Adv. Opt. Mater. **5**, 1700617 (2017)
104. Z. Wang, X. Quan, Z. Zhang, P. Cheng, Optical absorption of carbon-gold core-shell nanoparticles. J. Quant. Spectrosc. Radiat. Transf. **205**, 291–298 (2018)
105. M. Ha, J.H. Kim, M. You, Q. Li, C. Fan, J.M. Nam, Multicomponent plasmonic nanoparticles: from heterostructured nanoparticles to colloidal composite nanostructures. Chem. Rev. **119**, 12208–12278 (2019)
106. E. Prodan, C. Radloff, N.J. Halas, P. Nordlander, A hybridization model for the plasmon response of complex nanostructures. Science **302**, 419–422 (2003)
107. I. Jung, J. Kim, S. Lee, W. Park, S. Park, Multiple stepwise synthetic pathways toward complex plasmonic 2D and 3D nanoframes for generation of electromagnetic hot zones in a single entity. Acc. Chem. Res. **56**, 270–283 (2023)
108. X. Huang, S. Tang, B. Liu, B. Ren, N. Zheng, Enhancing the photothermal stability of plasmonic metal nanoplates by a core-shell architecture. Adv. Mater. **23**, 3420–3425 (2011)
109. A. Alkurdi, J. Lombard, F. Detcheverry, S. Merabia, Enhanced heat transfer with metal-dielectric core-shell nanoparticles. Phys. Rev. Appl. **13** (2020)
110. J. Cunha, T.L. Guo, G. Della Valle, A.N. Koya, R. Proietti Zaccaria, A. Alabastri, Controlling light, heat, and vibrations in plasmonics and phononics. Adv. Opt. Mater. **8** (2020) 2001225
111. C. Rossner, T.A.F. König, A. Fery, Plasmonic properties of colloidal assemblies. Adv. Opt. Mater. **9**, 2001869 (2021)
112. L. Lermusiaux, V. Many, P. Barois, V. Ponsinet, S. Ravaine, E. Duguet, M. Treguer-Delapierre, A. Baron, Toward Huygens' sources with dodecahedral plasmonic clusters. Nano Lett. **21**, 2046–2052 (2021)
113. L.O. Herrmann, V.K. Valev, C. Tserkezis, J.S. Barnard, S. Kasera, O.A. Scherman, J. Aizpurua, J.J. Baumberg, Threading plasmonic nanoparticle strings with light. Nat. Commun. **5**, 4568 (2014)
114. V. Myroshnychenko, J. Rodríguez-Fernández, I. Pastoriza-Santos, A.M. Funston, C. Novo, P. Mulvaney, L.M. Liz-Marzán, F.J. García de Abajo, Modelling the optical response of gold nanoparticles. Chem. Soc. Rev. **37**, 1792 (2008)

115. K.H. Su, Q.H. Wei, X. Zhang, J.J. Mock, D.R. Smith, S. Schultz, Interparticle coupling effects on plasmon resonances of nanogold particles. Nano Lett. **3**, 1087–1090 (2003)
116. D. Liu, C. Xue, Plasmonic coupling architectures for enhanced photocatalysis. Adv. Mater. **e2005738** (2021)
117. I. Romero, J. Aizpurua, G.W. Bryant, F.J. García De Abajo, Plasmons in nearly touching metallic nanoparticles: singular response in the limit of touching dimers. Opt. Express **14**, 9988 (2006)
118. P. Tao, G. Ni, C. Song, W. Shang, J. Wu, J. Zhu, G. Chen, T. Deng, Solar-driven interfacial evaporation. Nat. Energy **3**, 1031–1041 (2018)
119. A. Sanchot, G. Baffou, R. Marty, A. Arbouet, R. Quidant, C. Girard, E. Dujardin, Plasmonic nanoparticle networks for light and heat concentration. ACS Nano **6**, 3434–3440 (2012)
120. L. Zhou, Y. Tan, J. Wang, W. Xu, Y. Yuan, W. Cai, S. Zhu, J. Zhu, 3D self-assembly of aluminium nanoparticles for plasmon-enhanced solar desalination. Nat. Photon. **10**, 393–398 (2016)
121. Y. Liu, S. Yu, R. Feng, A. Bernard, Y. Liu, Y. Zhang, H. Duan, W. Shang, P. Tao, C. Song, T. Deng, A bioinspired, reusable, paper-based system for high-performance large-scale evaporation. Adv. Mater. **27**, 2768–2774 (2015)
122. M. Zhu, Y. Li, F. Chen, X. Zhu, J. Dai, Y. Li, Z. Yang, X. Yan, J. Song, Y. Wang, E. Hitz, W. Luo, M. Lu, B. Yang, L. Hu, Plasmonic wood for high-efficiency solar steam generation. Adv. Energy Mater. **8**, 1701028 (2018)
123. L. Tian, J. Luan, K.K. Liu, Q. Jiang, S. Tadepalli, M.K. Gupta, R.R. Naik, S. Singamaneni, Plasmonic biofoam: a versatile optically active material. Nano Lett. **16**, 609–616 (2016)
124. L. Zhou, Y. Tan, D. Ji, B. Zhu, P. Zhang, J. Xu, Q. Gan, Z. Yu, J. Zhu, Self-assembly of highly efficient, broadband plasmonic absorbers for solar steam generation. Sci. Adv. **2**, e1501227 (2016)
125. C. Kuppe, K.R. Rusimova, L. Ohnoutek, D. Slavov, V.K. Valev, "Hot" in plasmonics: temperature-related concepts and applications of metal nanostructures. Adv. Opt. Mater. **8**, 1901166 (2019)
126. J.A. Bordley, N. Hooshmand, M.A. El-Sayed, The coupling between gold or silver nanocubes in their homo-dimers: a new coupling mechanism at short separation distances. Nano Lett. **15**, 3391–3397 (2015)
127. J.J. Baumberg, J. Aizpurua, M.H. Mikkelsen, D.R. Smith, Extreme nanophotonics from ultrathin metallic gaps. Nat. Mater. **18**, 668–678 (2019)
128. D. Yoo, F. de León-Pérez, M. Pelton, I.H. Lee, D.A. Mohr, M.B. Raschke, J.D. Caldwell, L. Martín-Moreno, S.H. Oh, Ultrastrong plasmon-phonon coupling via epsilon-near-zero nanocavities. Nat. Photon. **15**, 125–130 (2020)
129. G. Baffou, R. Quidant, F.J. Garcia de Abajo, Nanoscale control of optical heating in complex plasmonic systems. ACS Nano **4**, 709–716 (2010)
130. K. Setoura, Y. Okada, D. Werner, S. Hashimoto, Observation of nanoscale cooling effects by substrates and the surrounding media for single gold nanoparticles under CW-laser illumination. ACS Nano **7**, 7874–7885 (2013)
131. Z. Qian, D.S. Ginger, Reversibly reconfigurable colloidal plasmonic nanomaterials. J. Am. Chem. Soc. **139**, 5266–5276 (2017)
132. S. Homaeigohar, M. Elbahri, Switchable plasmonic nanocomposites. Adv. Opt. Mater. **7** (2018)
133. X. Duan, N. Liu, Magnesium for dynamic nanoplasmonics. Acc. Chem. Res. **52**, 1979–1989 (2019)
134. S. Lee, K. Sim, S.Y. Moon, J. Choi, Y. Jeon, J.M. Nam, S.J. Park, Controlled assembly of plasmonic nanoparticles: from static to dynamic nanostructures. Adv. Mater. e2007668 (2021)
135. E.B.L.D. Sio, Active plasmonic nanomaterials CRC Press © 2016 by Taylor & Francis Group, LLC.
136. Z. Li, Y. Yin, Stimuli-responsive optical nanomaterials. Adv. Mater. **31**, 1807061 (2019)
137. L. Lin, X. Peng, M. Wang, L. Scarabelli, Z. Mao, L.M. Liz-Marzan, M.F. Becker, Y. Zheng, Light-directed reversible assembly of plasmonic nanoparticles using plasmon-enhanced thermophoresis. ACS Nano **10**, 9659–9668 (2016)

138. Y. Montelongo, D. Sikdar, Y. Ma, A.J.S. McIntosh, L. Velleman, A.R. Kucernak, J.B. Edel, A.A. Kornyshev, Electrotunable nanoplasmonic liquid mirror. Nat. Mater. **16**, 1127–1135 (2017)
139. S. Khatua, W.S. Chang, P. Swanglap, J. Olson, S. Link, Active modulation of nanorod plasmons. Nano Lett. **11**, 3797–3802 (2011)
140. M.S. Ergoktas, G. Bakan, E. Kovalska, L.W. Le Fevre, R.P. Fields, P. Steiner, X. Yu, O. Salihoglu, S. Balci, V.I. Fal'ko, K.S. Novoselov, R.A.W. Dryfe, C. Kocabas, Multispectral graphene-based electro-optical surfaces with reversible tunability from visible to microwave wavelengths. Nat. Photon. **15**, 493–498 (2021)
141. A. Kuzyk, R. Schreiber, H. Zhang, A.O. Govorov, T. Liedl, N. Liu, Reconfigurable 3D plasmonic metamolecules. Nat. Mater. **13**, 862–866 (2014)
142. L. Liu, R. Aleisa, Y. Zhang, J. Feng, Y. Zheng, Y. Yin, W. Wang, Dynamic color-switching of plasmonic nanoparticle films. Angew. Chem. Int. Ed. **58**, 16307–16313 (2019)
143. T. Ding, J.J. Baumberg, Thermo-responsive plasmonic systems: old materials with new applications. Nanoscale Adv. **2**, 1410–1416 (2020)
144. W. Lewandowski, M. Fruhnert, J. Mieczkowski, C. Rockstuhl, E. Górecka, Dynamically self-assembled silver nanoparticles as a thermally tunable metamaterial. Nat. Commun. **6** (2015)
145. T. Ding, V.K. Valev, A.R. Salmon, C.J. Forman, S.K. Smoukov, O.A. Scherman, D. Frenkel, J.J. Baumberg, Light-induced actuating nanotransducers. Proc. Natl. Acad. Sci. **113**, 5503–5507 (2016)
146. L. Liu, Z. Gao, B. Jiang, Y. Bai, W. Wang, Y. Yin, Reversible assembly and dynamic plasmonic tuning of Ag nanoparticles enabled by limited ligand protection. Nano Lett. **18**, 5312–5318 (2018)
147. Z. Li, W. Wang, Y. Yin, Colloidal assembly and active tuning of coupled plasmonic nanospheres. Trends Chem. **2**, 593–608 (2020)
148. Z. Sun, W. Ni, Z. Yang, X. Kou, L. Li, J. Wang, PH-controlled reversible assembly and disassembly of gold nanorods. Small **4**, 1287–1292 (2008)
149. Y. Liu, X. Han, L. He, Y. Yin, Thermoresponsive assembly of charged gold nanoparticles and their reversible tuning of plasmon coupling. Angew. Chem. Int. Ed. **51**, 6373–6377 (2012)
150. G.E. Lio, G. Palermo, A. De Luca, R. Caputo, Tensile control of the thermal flow in plasmonic heaters realized on flexible substrates. J. Chem. Phys. **151**, 244707 (2019)
151. G.E. Lio, G. Palermo, R. Caputo, A. De Luca, Opto-mechanical control of flexible plasmonic materials. J. Appl. Phys. **125**, 082533 (2019)
152. G. Emanuele Lio, A. De Luca, C.P. Umeton, R. Caputo, Opto-mechanically induced thermo-plasmonic response of unclonable flexible tags with hotspot fingerprint. J. Appl. Phys. **128**, 093107 (2020)
153. N.C. Lindquist, P. Nagpal, K.M. McPeak, D.J. Norris, S.H. Oh, Engineering metallic nanostructures for plasmonics and nanophotonics, Reports on progress in physics. Phys. Soc. **75**, 036501 (2012)
154. S. Kasani, K. Curtin, N. Wu, A review of 2D and 3D plasmonic nanostructure array patterns: fabrication, light management and sensing applications. Nanophotonics **8**, 2065–2089 (2019)
155. T. Qiu, E.M. Akinoglu, B. Luo, M. Konarova, J.H. Yun, I.R. Gentle, L. Wang, Nanosphere lithography: a versatile approach to develop transparent conductive films for optoelectronic applications. Adv. Mater. **34**, e2103842 (2022)
156. L.J. Guo, Nanoimprint lithography: methods and material requirements. Adv. Mater. **19**, 495–513 (2007)
157. M. Campbell, D.N. Sharp, M.T. Harrison, R.G. Denning, A.J. Turberfield, Fabrication of photonic crystals for the visible spectrum by holographic lithography. Nature **404**, 53–56 (2000)
158. K. Yang, X. Yao, B. Liu, B. Ren, Metallic plasmonic array structures: principles, fabrications, properties, and applications. Adv. Mater. **33**, e2007988 (2021)
159. J. Fischer, M. Wegener, Three-dimensional optical laser lithography beyond the diffraction limit. Laser Photon. Rev. **7**, 22–44 (2012)

160. Y. Wang, M. Zhang, Y. Lai, L. Chi, Advanced colloidal lithography: From patterning to applications. Nano Today **22**, 36–61 (2018)
161. Z. Lu, M. Zhou, Fabrication of large scale two-dimensional colloidal crystal of polystyrene particles by an interfacial self-ordering process. J. Colloid Interface Sci. **361**, 429–435 (2011)
162. Y. Yu, C. Ng, T.A.F. Konig, A. Fery, Tackling the scalability challenge in plasmonics by wrinkle-assisted colloidal self-assembly. Langmuir ACS J. Surf. Colloids **35**, 8629–8645 (2019)
163. C. Zhu, G. Meng, P. Zheng, Q. Huang, Z. Li, X. Hu, X. Wang, Z. Huang, F. Li, N. Wu, A hierarchically ordered array of silver-nanorod bundles for surface-enhanced Raman scattering detection of phenolic pollutants. Adv. Mater. **28**, 4871–4876 (2016)
164. K. Xu, Z. Wang, C.F. Tan, N. Kang, L. Chen, L. Ren, E.S. Thian, G.W. Ho, R. Ji, M. Hong, Uniaxially stretched flexible surface plasmon resonance film for versatile surface enhanced Raman scattering diagnostics. ACS Appl. Mater. Interfaces **9**, 26341–26349 (2017)
165. S.M. Park, X.G. Liang, B.D. Harteneck, T.E. Pick, N. Hiroshiba, Y. Wu, B.A. Helms, D.L. Olynick, Sub-10 nm nanofabrication nanoimprint directed self-assembly of block copolymers. ACS Nano **5**, 8523–8531 (2011)
166. B.W. Liu, X. Yao, L. Zhang, H.X. Lin, S. Chen, J.H. Zhong, S. Liu, L. Wang, B. Ren, Efficient platform for flexible engineering of superradiant, fano-type, and subradiant resonances. ACS Photon. **2**, 1725–1731 (2015)
167. J. Park, K.I. Kim, K. Kim, D.C. Kim, D. Cho, J.H. Lee, S. Jeon, Rapid, high-resolution 3D interference printing of multilevel ultralong nanochannel arrays for high-throughput nanofluidic transport. Adv. Mater. **27**, 8000–8006 (2015)
168. X. Zhang, M. Theuring, Q. Song, W. Mao, M. Begliarbekov, S. Strauf, Holographic control of motive shape in plasmonic nanogap arrays. Nano Lett. **11**, 2715–2719 (2011)
169. L. Duempelmann, D. Casari, A. Luu-Dinh, B. Gallinet, L. Novotny, Color rendering plasmonic aluminum substrates with angular symmetry breaking. ACS Nano **9**, 12383–12391 (2015)
170. H.G. Duan, H.L. Hu, K. Kumar, Z.X. Shen, J.K.W. Yang, Direct and reliable patterning of plasmonic nanostructures with sub-10-nm gaps. ACS Nano **5**, 7593–7600 (2011)
171. M.K. Kim, H. Sim, S.J. Yoon, S.H. Gong, C.W. Ahn, Y.H. Cho, Y.H. Lee, Squeezing photons into a point-like space. Nano Lett. **15**, 4102–4107 (2015)
172. X. Zhu, C. Vannahme, E. Hojlund-Nielsen, N.A. Mortensen, A. Kristensen, Plasmonic colour laser printing. Nat. Nanotechnol. **11**, 325–329 (2016)
173. X. Zhu, J. Engelberg, S. Remennik, B. Zhou, J.N. Pedersen, P. Uhd Jepsen, U. Levy, A. Kristensen, Resonant laser printing of optical metasurfaces. Nano Lett. **22**, 2786–2792 (2022)
174. L. Chen, Y. Zhou, Y. Li, M. Hong, Microsphere enhanced optical imaging and patterning: From physics to applications. Appl. Phys. Rev. **6** (2019)
175. Q.C. Tong, D.T.T. Nguyen, M.T. Do, M.H. Luong, B. Journet, I. Ledoux-Rak, N.D. Lai, Direct laser writing of polymeric nanostructures via optically induced local thermal effect. Appl. Phys. Lett. **108** (2016)
176. L. Qin, Y. Huang, F. Xia, L. Wang, J. Ning, H. Chen, X. Wang, W. Zhang, Y. Peng, Q. Liu, Z. Zhang, 5 nm nanogap electrodes and arrays by super-resolution laser lithography. Nano Lett. **20**, 4916–4923 (2020)
177. Z. Lin, H. Liu, L. Ji, W. Lin, M. Hong, Realization of approximately 10 nm features on semiconductor surfaces via femtosecond laser direct patterning in far field and in ambient air. Nano Lett. **20**, 4947–4952 (2020)
178. H.C. Jeon, T.Y. Jeon, T.S. Shim, S.M. Yang, Direct fabrication of hexagonally ordered ridged nanoarchitectures via dual interference lithography for efficient sensing applications. Small **10**, 1490–1494 (2014)
179. B. Liu, X. Yao, S. Chen, H. Lin, Z. Yang, S. Liu, B. Ren, Large-area hybrid plasmonic optical cavity (HPOC) substrates for surface-enhanced Raman spectroscopy. Adv. Funct. Mater. **28** (2018)
180. C. Zhan, B.W. Liu, Y.F. Huang, S. Hu, B. Ren, M. Moskovits, Z.Q. Tian, Disentangling charge carrier from photothermal effects in plasmonic metal nanostructures. Nat. Commun. **10**, 2671 (2019)

181. B. Liu, C. Zhan, X. Yao, S. Yan, B. Ren, Nanobowtie arrays with tunable materials and geometries fabricated by holographic lithography. Nanoscale **12**, 21401–21408 (2020)
182. T.T. Tran, D. Wang, Z.Q. Xu, A. Yang, M. Toth, T.W. Odom, I. Aharonovich, Deterministic coupling of quantum emitters in 2D materials to plasmonic nanocavity arrays. Nano Lett. **17**, 2634–2639 (2017)
183. B. Lee, J. Park, G.H. Han, H.S. Ee, C.H. Naylor, W. Liu, A.T. Johnson, R. Agarwal, Fano resonance and spectrally modified photoluminescence enhancement in monolayer MoS2 integrated with plasmonic nanoantenna array. Nano Lett. **15**, 3646–3653 (2015)
184. P. Yang, J. Zheng, Y. Xu, Q. Zhang, L. Jiang, Colloidal synthesis and applications of plasmonic metal nanoparticles. Adv. Mater. **28**, 10508–10517 (2016)
185. Q.N. Nguyen, C. Wang, Y. Shang, A. Janssen, Y. Xia, Colloidal synthesis of metal nanocrystals: from asymmetrical growth to symmetry breaking. Chem. Rev. **123**, 3693–3760 (2023)
186. T. Hueckel, G.M. Hocky, S. Sacanna, Total synthesis of colloidal matter. Nat. Rev. Mater. **6**, 1053–1069 (2021)
187. G. Schmid, L.F. Chi, Metal clusters and colloids. Adv. Mater. **10**, 515–526 (1998)
188. P.S.J. Turkevich, J. Hillier, A study of the nucleation and growth processes in the synthesis of colloidal gold. Discuss. Faraday Soc. 55–75 (1951)
189. G.B. Matthew Pelton, *Introduction to Metal-Nanoparticle Plasmonics* (Wiley, 2013), p 275
190. M. Brust, M. Walker, D. Bethell, D.J. Schiffrin, R. Whyman, Synthesis of thiol-derivatised gold nanoparticles in a two-phase Liquid–Liquid system. J. Chem. Soc. Chem. Commun. 801–802 (1994)
191. S. Shrestha, B. Wang, P. Dutta, Nanoparticle processing: understanding and controlling aggregation. Adv. Coll. Interface. Sci. **279**, 102162 (2020)
192. Y. Xia, K.D. Gilroy, H.C. Peng, X. Xia, Seed-mediated growth of colloidal metal nanocrystals. Angew. Chem. **56**, 60–95 (2017)
193. S.E. Skrabalak, Y. Xia, Pushing nanocrystal synthesis toward nanomanufacturing. ACS Nano **3**, 10–15 (2009)
194. S. Gwo, H.Y. Chen, M.H. Lin, L. Sun, X. Li, Nanomanipulation and controlled self-assembly of metal nanoparticles and nanocrystals for plasmonics. Chem. Soc. Rev. **45**, 5672–5716 (2016)
195. M. Mayer, M.J. Schnepf, T.A.F. König, A. Fery, Colloidal self-assembly concepts for plasmonic metasurfaces. Adv. Opt. Mater. **7** (2018)
196. O.N. Oliveira Jr., L. Caseli, K. Ariga, The past and the future of langmuir and langmuir-blodgett films. Chem. Rev. **122**, 6459–6513 (2022)
197. J.A. Fan, C. Wu, K. Bao, J. Bao, R. Bardhan, N.J. Halas, V.N. Manoharan, P. Nordlander, G. Shvets, F. Capasso, Self-assembled plasmonic nanoparticle clusters. Science **328**, 1135–1138 (2010)
198. B. Tim, M. Kotkowiak, N. Kowalska, A.B. Nowicka, W. Lewandowski, Influence of gold nanoparticle assembly in langmuir-schaefer monolayers on the surface-enhanced spectroscopy response of a nanoplatform. J. Phys. Chem. C **127**, 15978–15987 (2023)
199. L. Scarabelli, D. Vila-Liarte, A. Mihi, L.M. Liz-Marzán, Templated colloidal self-assembly for lattice plasmon engineering. Accounts Mater. Res. **2**, 816–827 (2021)
200. M.R. Jones, K.D. Osberg, R.J. Macfarlane, M.R. Langille, C.A. Mirkin, Templated techniques for the synthesis and assembly of plasmonic nanostructures. Chem. Rev. **111**, 3736–3827 (2011)
201. B.J. Jankiewicz, D. Jamiola, J. Choma, M. Jaroniec, Silica-metal core-shell nanostructures. Adv. Coll. Interface. Sci. **170**, 28–47 (2012)
202. Q. Xu, G. Meng, F. Han, Porous AAO template-assisted rational synthesis of large-scale 1D hybrid and hierarchically branched nanoarchitectures. Prog. Mater. Sci. **95**, 243–285 (2018)
203. K. Wang, G. Liu, N. Hoivik, E. Johannessen, H. Jakobsen, Electrochemical engineering of hollow nanoarchitectures: pulse/step anodization (Si, Al, Ti) and their applications. Chem. Soc. Rev. **43**, 1476–1500 (2014)

204. W. Lee, S.J. Park, Porous anodic aluminum oxide: anodization and templated synthesis of functional nanostructures. Chem. Rev. **114**, 7487–7556 (2014)
205. Z. Liu, N. Liu, J. Schroers, Nanofabrication through molding. Prog. Mater. Sci. **125**, 100891 (2022)
206. V. Flauraud, M. Mastrangeli, G.D. Bernasconi, J. Butet, D.T. Alexander, E. Shahrabi, O.J. Martin, J. Brugger, Nanoscale topographical control of capillary assembly of nanoparticles. Nat. Nanotechnol. **12**, 73–80 (2017)
207. Y. Yin, Y. Lu, B. Gates, Y. Xia, Template-assisted self-assembly: a practical route to complex aggregates of monodispersed colloids with well-defined sizes, shapes, and structures. J. Am. Chem. Soc. **123**, 8718–8729 (2001)

# Chapter 4
# Characterization Techniques

The ability to accurately measure localized temperature variations is essential for comprehending the fundamental aspects of thermoplasmonics. This calls for nanothermometry that possesses high spatial, temporal, and thermal resolutions. Yet, the sensitivity of many current techniques is compromised by systematic errors, which can arise from fluctuations in fluorescence, variations in local environmental conditions, or changes in the optical properties of the surrounding medium. Therefore, it is crucial to combine analytical models, numerical simulations, and experimental methods to achieve a complement understanding of thermoplasmonics.

## 4.1 Analytical Model

Analytical solutions to Maxwell's equations are limited to a few specific geometries, such as the application of Mie's theory for spheres [1]. For anisotropic particles in the long-wavelength limit, Gans' theory is commonly used to model the electrostatic response of ellipsoids [2]. The collective optical behavior of particle arrays is explored through analytical multiple scattering theories [3], and semi-analytical modal expansions offer a detailed description of both near and far fields [4]. It is crucial to note that for any given geometry, plasmon resonance wavelengths and quality factors are closely linked to the resonant values of the complex dielectric permittivity of the material [5]. In this section, we utilize a comprehensive solution of Maxwell's equations that involves successive retardation corrections within a perturbative expansion framework. This method enables us to accurately compute extinction and scattering cross sections, alongside plasmon frequencies and lifetimes, using just a small number of real parameters.



G. Liu, *Thermoplasmonics*, https://doi.org/10.1007/978-981-97-8332-8_4

### 4.1.1 Electrostatic Limit

We analyze the scenario where illumination is provided by monochromatic light at a frequency of $\omega$, leading to a time-dependent electric field expressed as $\mathbf{E}(\mathbf{r}, t) = \mathbf{E}(\mathbf{r}, \omega)e^{-i\omega t} + \mathbf{E}^*(\mathbf{r}, \omega)e^{i\omega t}$. We examine homogeneous metallic particles, characterizing the metal via its frequency-dependent permittivity $\varepsilon_m(\omega)$. This approach is known as the local approximation, implying that the dielectric displacement, $\varepsilon_m(\omega)\mathbf{E}(\mathbf{r}, \omega)$, solely depends on the electric field at the identical spatial position $\mathbf{r}$. However, this approximation tends to fail for very small particles—typically less than 10 nm for noble metals and in cases involving either pronounced sharp tips or very narrow gaps between metals (less than 1 nm) [6].

When dealing with small particles relative to the light wavelength, retardation effects become insignificant, and the optical response can be modeled by solving Poisson's equation, $\nabla \cdot \varepsilon(\mathbf{r}, \omega)\mathbf{E}(\mathbf{r}, \omega) = 0$. Here, $\varepsilon(\mathbf{r}, \omega) = \varepsilon_m(\omega) + (1 - f(\mathbf{r}))\varepsilon_h(\omega)$, where $\varepsilon_h(\omega)$ is the permittivity of the host dielectric, and $f(\mathbf{r})$ is a filling function defined as 1 inside the particle and 0 elsewhere. For a significant duration, modal expansions have provided semi-analytical frameworks to address this electrostatic scenario [7, 8]. These expansions fundamentally depend on a set of orthogonal and complete real eigenmodes $\mathbf{E}_j(\mathbf{r})$ and eigenvalues $\varepsilon_j$, indexed by $j$, which are consistent within the subspace of electrostatic solutions:

$$\nabla \cdot \left[\varepsilon_j f(\mathbf{r}) + (1 - f(\mathbf{r})\right]\mathbf{E}_j(\mathbf{r}) = 0 \tag{4.1}$$

$\mathbf{E}_j$ represents the independent electrostatic field generated by a particle with identical geometry situated in a vacuum and infused with a medium of permittivity $\varepsilon_j$. These eigenmodes are normalized to adhere to an orthogonality relation.

$$\int d^3\mathbf{r} f(\mathbf{r})\mathbf{E}_j(\mathbf{r}) \cdot \mathbf{E}'_j(\mathbf{r}) = L^3\delta'_{jj} \tag{4.2}$$

where $L$ represents a characteristic length of the particle, such as the diameter in the case of a sphere or the elongation for anisotropic particles. Notably, these modes are unaffected by the material composition of the particle. However, they enable us to articulate the self-consistent electric field $\mathbf{E}(\mathbf{r}, \omega)$ in the vicinity of an actual nanoparticle with permittivity $\varepsilon_m(\omega)$, responding to an externally applied field $\mathbf{E}^{\text{ext}}(\mathbf{r}, \omega)$, described as:

$$\mathbf{E}(\mathbf{r}, \omega) = \sum_j \left[1 - \frac{\frac{\varepsilon_m}{\varepsilon_h} - 1}{\varepsilon_j - 1}\right]^{-1} C_j^{\text{ext}}(\omega)\mathbf{E}_j(\mathbf{r}) \tag{4.3}$$

where $\varepsilon_h(\omega)$ is the real, positive permittivity of the host medium, and

$$C_j^{\text{ext}}(\omega) = \frac{1}{L^3}\int d^3\mathbf{r} f(\mathbf{r})\mathbf{E}_j(\mathbf{r}) \cdot \mathbf{E}^{\text{ext}}(\mathbf{r}, \omega) \tag{4.4}$$

The particle polarizability along a symmetry direction can also be expanded to include contributions from various eigenmodes, expressed as:

$$\alpha(\omega) = \frac{\varepsilon_h}{4\pi}\sum_j V_j\left(\frac{1}{\frac{\varepsilon_m}{\varepsilon_h}-1}-\frac{1}{\varepsilon_j-1}\right)^{-1} \tag{4.5}$$

The mode volumes $V_j$ are defined as $(1/L^3)\left|\int d^3\mathbf{r}f(\mathbf{r})\mathbf{E}_j(\mathbf{r})\right|^2$, and the sum of these volumes equals the total volume of the particle, $\sum_j V_j = V$, courtesy of the noted orthonormality. Specifically, for each symmetry axis $l = x, y, z$ of an ellipsoid, there exists only one electrostatic dipolar mode. Consequently, there is just one term $(j = 1)$ in the equation above, with $V_1$ equating to $V$ and $\varepsilon_1 = 1 - 1/L_l$ where:

$$L_l = (a_x a_y a_z/2)\int_0^\infty ds\left(s+a_l^2\right)^{-1}\left[\left(s+a_x^2\right)\left(s+a_y^2\right)\left(s+a_z^2\right)\right]^{-1/2} \tag{4.6}$$

is the corresponding depolarization factor, with $a_{xyz}$ representing the half-axis lengths [2].

### 4.1.2 Retardation Corrections

We extended the electrostatic limit to consider the effects arising from the particle's finite size relative to the wavelength of light in free space, $\lambda$. Specifically, we use the size factor $s = \sqrt{\varepsilon_h}L/\lambda$ as a useful perturbation parameter, which inherently adjusts for the decreased wavelength in the host medium compared to free space, an effect attributed to the refractive index $\sqrt{\varepsilon_h}$. Consequently, the polarizability described in Eq. (4.5) has been revised as [2]

$$\alpha(\omega) = \frac{\varepsilon_h}{4\pi}\sum_j V_j B_j(s)\left(\frac{1}{\frac{\varepsilon_m}{\varepsilon_h}-1}-\frac{1}{\varepsilon_j-1}-A_j(s)\right)^{-1} \tag{4.7}$$

In this context, $A_j(s)$ and $B_j(s)$ are functions that depend on $s$. Practically, we focus on scenarios where $s$ is relatively small, prompting a natural expansion of these functions into power series of $s$. We observe that plasmon resonances undergo shifts and broadening, which can be attributed to the nonzero real and imaginary components in $A_j$, respectively. From this observation, we derive the following expansion:

$$A_j = a_{j2}s^2 + \frac{4\pi^2 iV_j}{3L^3}s^3 + a_{j4}s^4 + \dots \tag{4.8}$$

where

$$a_{jn} = \frac{(2i)^n}{4n(n-2)!L^{n+3}} \int d^3 r f(\mathbf{r}) \int d^3 r' f(\mathbf{r}')$$
$$\times \left\{ (n-3) |\mathbf{r} - \mathbf{r}'|^{n-5} [(\mathbf{r} - \mathbf{r}') \cdot \mathbf{E}_j(\mathbf{r})] [(\mathbf{r} - \mathbf{r}') \cdot \mathbf{E}_j(\mathbf{r}')] \right.$$
$$\left. + (1-n) |\mathbf{r} - \mathbf{r}'|^{n-3} \mathbf{E}_j(\mathbf{r}) \cdot \mathbf{E}_j(\mathbf{r}') \right\} \tag{4.9}$$

Furthermore, considering the range of $s$ being evaluated, the correction in resonance strengths (as indicated by the numerator of Eq. (4.7)) is small, making it reasonable to approximate $B_j \approx 1$. Similarly, despite the mix of electrostatic modes $\mathbf{E}_j$ in the electric field of Eq. (4.3) caused by retardation, this effect appears minimal for the $s$ values under consideration. The main correction emerges from frequency shifts and broadening, encapsulated by the following expression

$$\mathbf{E}(\mathbf{r}, \omega) \approx \sum_j \left[ 1 - \left( \frac{\varepsilon_m}{\varepsilon_h} - 1 \right) \left( \frac{1}{\varepsilon_j - 1} + A_j \right) \right]^{-1} C_j^{\text{ext}}(\omega) \mathbf{E}_j(\mathbf{r}) \tag{4.10}$$

This method can effectively explain the far-field scattering (via $\alpha$ in Eq. (4.7)) and near-field response (via $\mathbf{E}$ in Eq. (4.10)) involving metal nanoparticles that factor in retardation, utilizing the electrostatic modes s $\mathbf{E}_j$ and a set of four real numbers comprising $\varepsilon_j$, $V_j$, $a_{j2}$, and $a_{j4}$. By fitting the position and strength of the peak connected with mode $j$ in the absorption spectrum of the particle—calculated under the electrostatic limit—the resonant permittivity $\varepsilon_j$ is ascertained. The near field $\mathbf{E}_j$, meanwhile, is considered as the induced field computed when the light frequency matches the absorption maximum. After normalizing the field according to Eq. (4.2), mode volume $V_j$ is then derived, along with $a_{j2}$ and $a_{j4}$ from Eq. (4.9).

In recent advances, a unique set of quasi-normal modes has been deployed to deal with ohmic and radiative losses for plasmonic particles [4, 9, 10]. These modes, which are derived from the Maxwell equations, are contingent on the metal's dielectric function. Their normalization also requires careful consideration, such as in Purcell factors' calculation [9]. In contrast, the eigenmodes of this methodology are free from the influence of metal permittivity selection and only involve the particle volume in their normalization process.

### 4.1.3 Plasmon Frequency

Plasmons correlate with resonant conditions of the metal's dielectric function nearing one of the eigenvalues $\varepsilon_j$, which are negative real numbers. Actual metals showcase losses proportional to $\text{Im}(\varepsilon_m) > 0$, denoting finite plasmon lifetimes. It is practical to approximate $\varepsilon_m(\omega) \approx \varepsilon_m(\omega_j) + \varepsilon'_m(\omega_j)(\omega - \omega_j)$ near the real frequency $\omega_j$, where $\text{Re}\{\varepsilon_m(\omega_j)\} = \varepsilon_h \varepsilon_j$ and the prime signifies differentiation with respect to the variable. In scenarios devoid of retardation, the complex plasmon frequency $\tilde{\omega}_j$ must meet the condition $\varepsilon_m(\tilde{\omega}_j) = \varepsilon_h \varepsilon_j$. Considering this approximation and assuming $\mathrm{Im}(\varepsilon_m) \ll |\varepsilon_m|$, it leads to $\tilde{\omega}_j \approx \omega_j - i/2\tau_j$, where $\tau_j = \mathrm{Re}\{\varepsilon'_m(\omega_j)\}/\mathrm{Im}\{\varepsilon_m(\omega_j)\}$. From the physical prerequisite $\mathrm{Re}\{\varepsilon_m + \omega\varepsilon'_m\} \geq 0$, we deduce $\tau_j > 0$. After stopping of external excitation, the near-field intensity diminishes over time proportionally to $\propto \left|e^{-i\tilde{\omega}_j t}\right|^2 = e^{-t/\tau_j}$ distinctly marking $\tau_j$ as the plasmon lifetime [2].

In common materials, noble metals exhibit the most favorable small values of $\mathrm{Im}(\varepsilon_m)$ particularly within the visible and near-infrared spectrum. Notably, gold, silver, and copper display a Drude-like response as given by the formula:

$$\varepsilon_m(\omega) = \varepsilon_b - \frac{\omega_p^2}{\omega\left(\omega + i\tau^{-1}\right)} \tag{4.11}$$

Here, the parameters $\varepsilon_b$, $\omega_p$, and $\tau$ as detailed in Table 4.1 [2]. This model effectively describes the measured dielectric function, using constant values of $\varepsilon_b$ at wavelengths exceeding the interband transition thresholds. These transitions enhance the polarization component in the real part of $\varepsilon_m$ (the $\varepsilon_b$ term of Eq. (4.11)) and also escalate the imaginary part (indicative of actual excitations) at shorter wavelengths, specifically below ~ 370 nm for Ag, ~ 500 nm for Au, and ~ 550 nm for Cu. Within these spectral regions, Eq. (4.11) remains applicable by adjusting the $\omega$-dependence in $\varepsilon_b$, as illustrated in Table 4.1 (refer to caption). At longer wavelengths, the lifetime of nonretarded plasmons is directly derived from the depicted Drude model values ($\tau_j = \tau$).

Retardation impacts the calculation through the term $A_j(s)$ Eq. (4.8), which, when considering small size parameters s (i.e., $|A_j| \ll 1$), and using Eq. (4.11), refines the calculation of the corresponding plasmon pole in Eq. (4.7). This results in an approximation for the complex plasmon frequency:

$$\tilde{\omega}_j \approx \frac{\omega_p}{\sqrt{\varepsilon_b - \varepsilon_h \varepsilon_j}}\left[1 - \frac{A_j}{2}\frac{\varepsilon_h\left(\varepsilon_j - 1\right)^2}{\varepsilon_b - \varepsilon_h \varepsilon_j}\right] - i/2\tau \tag{4.12}$$

**Table 4.1** Parameters associated with the Drude dielectric function either maintain a constant for $\varepsilon_b$ or substitute this constant with an $\omega$-dependent analytical fitting function $\varepsilon_b(\omega) = A + B\log\left[\left(\omega_1 - \omega - i\tau_1^{-1}\right)/\left(\omega_1 + \omega + i\tau_1^{-1}\right)\right] + C\exp(-\omega/\omega_2)$ for gold, silver, and copper [2]

| Material | $\varepsilon_b$ | $\hbar\omega_p$ (ev) | $\hbar\tau^{-1}$ (ev) | $\tau$ (fs) | $A$ | $B$ | $C$ | $\hbar\omega_1$ (ev) | $\hbar\tau_1^{-1}$ (ev) | $\hbar\omega_2$ (ev) |
|---|---|---|---|---|---|---|---|---|---|---|
| Au | 9.5 | 9.06 | 0.071 | 9.3 | 0.132 | –1.755 | 20.43 | 2.43 | 0.0716 | 1.52 |
| Ag | 4.0 | 9.17 | 0.021 | 31 | –9.71 | –1.111 | 13.77 | 4.02 | 0.0760 | 18.5 |
| Cu | 8.0 | 8.88 | 0.103 | 6.4 | –4.36 | –1.655 | 12.31 | 2.12 | 0.0528 | 5.43 |

This equation indicates a frequency red shift, primarily due to the positive $a_{j2}$ term in Eq. (4.8), which is generally proportional to $-(L/\lambda)^2$. This shift is paired with an increase in plasmon linewidth from $1/\tau$ to:

$$\Delta\omega_j \approx \frac{1}{\tau} + \frac{4\pi^2\omega_p}{3}\frac{V_j}{\lambda^3}\frac{{\varepsilon_h}^{5/2}(\varepsilon_j - 1)^2}{(\varepsilon_b - \varepsilon_h\varepsilon_j)^{3/2}} \tag{4.13}$$

in which the second term accounts for the effects of radiative damping.

## 4.2 Computational Approaches

To understand the optical behavior of intricate structures, various numerical methods are developed to handle arbitrary shapes [11, 12]. Among these, the most used techniques include the finite-difference time-domain (FDTD) method [13, 14], the discrete dipole approximation (DDA) [15, 16], Green's function approaches akin to DDA [17, 18], the multiple multipole (MMP) method [19], multiple scattering techniques, transfer-matrix methods [20], plane-wave expansions [21], and boundary element methods (BEM) [22, 23]. Each approach offers unique advantages and challenges. Plane-wave expansions and transfer-matrix methods are particularly effective when a limited set of expansion functions can precisely represent a structure's fields. This is often the case for objects of extended wavelength scale and for NP arrays. However, these methods become more challenging when applied to small, nanometer-scale sub-wavelength individual NPs with complex geometries, or for closely spaced, strongly coupled NPs, as well as structures incorporating multiple length scales [24]. Consequently, real-space techniques are commonly favored to analyze both the near-field and far-field responses of complex metallic structures.

### 4.2.1 *FDTD Method*

The FDTD method offers a direct approach by explicitly solving Maxwell's equations. Fields are discretized both in time and space across a grid that includes the nanoparticles under investigation and an adjacent region for field evaluation. The fields are calculated by advancing discrete versions of Maxwell's equations through time. For example, to understand this process, consider one of Maxwell's curl equations:

$$\nabla \times \boldsymbol{E} = -\frac{1}{c}\frac{\partial \boldsymbol{B}}{\partial t} \tag{4.14}$$

here $\boldsymbol{B}$ represents the magnetic induction. By applying a central difference formula for the time derivative, we can express this as:

$$\boldsymbol{B}(\mathbf{x}, t+\delta t) = \boldsymbol{B}(\mathbf{x}, t-\delta t) - 2(\delta t)c\nabla \times \boldsymbol{E}(\mathbf{x}, t) \tag{4.15}$$

Given $\boldsymbol{B}$ at time $(t - \delta t)$ and $\boldsymbol{E}$ at time $t$, $\boldsymbol{B}$ at time $(t + \delta t)$ is derived from this equation. Similarly, Maxwell's other curl equation facilitates the forward time propagation of $\boldsymbol{E}$. As a result, $\boldsymbol{E}$ and $\boldsymbol{B}$ are defined on two staggered time grids with a step size of $2\delta t$, shifted by $\delta t$. Likewise, the curl operator is represented as a sum of central differences, thus positioning $\boldsymbol{E}$ and $\boldsymbol{B}$ on staggered spatial grids. Once the initial conditions of the fields are set, propagating these values forward in time becomes a computationally straightforward task.

To prevent reflections from the computational region's edges, absorbing boundary conditions are typically used. These conditions allow outward-moving fields to exit without returning to the simulation area. The selection of initial fields hinges on the specific properties under investigation. Often, the objective is to evaluate the local fields within and surrounding a nanoparticle when it interacts with a uniform light wave. In such cases, the initial fields resemble a plane-wave pulse originating from a significant distance away from the nanoparticle. This field continues for a period much longer than required for the pulse to pass through the nanostructure and reach the boundaries of the computational region. To explore characteristics at specific frequencies, one can either run multiple simulations with monochromatic incident fields at different frequencies or use a broad-spectrum pulse and conduct Fourier analysis on the collected data.

The immediate outcome of such computations is the near-field distribution around the metal nanoparticle. Far-field properties, such as scattering and absorption cross sections, can be derived from these near fields through appropriate surface integrals. The total electric field around the nanoparticle can be described as $\boldsymbol{E} = \boldsymbol{E}_i + \boldsymbol{E}_s$, where $\boldsymbol{E}_i$ is the incident plane-wave field, and $\boldsymbol{E}_s$ is the scattered field. The total Poynting vector is given by:

$$\begin{aligned} \boldsymbol{S} &= \frac{c}{8\pi}\mathrm{Re}\{\boldsymbol{E} \times \boldsymbol{B}^*\} = \frac{c}{8\pi}\mathrm{Re}\{\boldsymbol{E}_i \times \boldsymbol{B}_i^*\} + \frac{c}{8\pi}\mathrm{Re}\{\boldsymbol{E}_s \times \boldsymbol{B}_s^*\} \\ &+ \frac{c}{8\pi}\mathrm{Re}\{\boldsymbol{E}_i \times \boldsymbol{B}_s^* + \boldsymbol{E}_s \times \boldsymbol{B}_i^*\} = \boldsymbol{S}_i + \boldsymbol{S}_s + \boldsymbol{S}_{ext} \end{aligned} \tag{4.16}$$

Here, $\boldsymbol{S}_i$ represents the energy flux of the incident field, $\boldsymbol{S}_s$ is the energy flux of the scattered field, and $\boldsymbol{S}_{ext}$ is the additional energy flux resulting from the interference terms. In a nonabsorbing medium, the energy absorbed by the nanoparticle $\boldsymbol{F}_a$ is the flux of energy through any surface $\Omega$ that surrounds the nanoparticle (Fig. 4.1):

$$\boldsymbol{F}_a = -\int \boldsymbol{S} \cdot \boldsymbol{n}\,\mathrm{d}\Omega \tag{4.17}$$

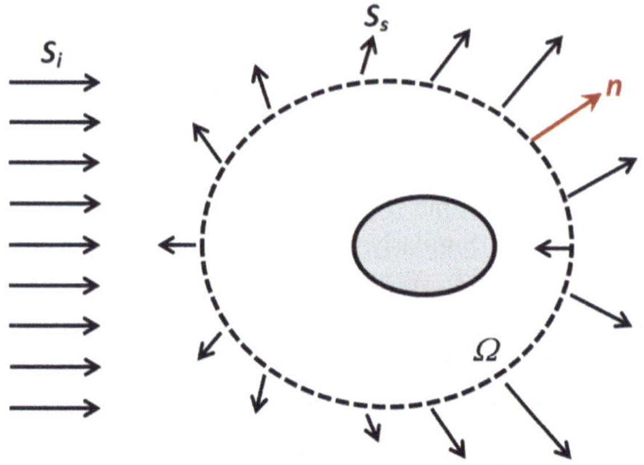


**Fig. 4.1** Energy flux $S_i$ of an incident field crossing an surface $\Omega$, with the resulting flux $S_s$ that scatters from the nanoparticle

here $\boldsymbol{n}$ represents the outward surface normal. The outward energy flow due to scattering is expressed as:

$$\boldsymbol{F}_s = -\int \boldsymbol{S}_s \cdot \boldsymbol{n} \mathrm{d}\Omega \tag{4.18}$$

The inward energy flow from the interference terms is given by:

$$\boldsymbol{F}_{ext} = -\int \boldsymbol{S}_{ext} \cdot \boldsymbol{n} \mathrm{d}\Omega \tag{4.19}$$

The energy flow of the incident plane wave across any surface ultimately vanishes from energy conservation, which yields:

$$\boldsymbol{F}_{ext} = \boldsymbol{F}_s + \boldsymbol{F}_{abs} \tag{4.20}$$

Here, $\boldsymbol{F}_{ext}$ equals the energy lost from the incident beam, accounted for by absorption into the nanoparticle and scattering. This phenomenon, involving both absorption and scattering, is termed extinction. Cross sections for extinction ($C_{\text{ext}}$), absorption ($C_{\text{abs}}$), and scattering ($C_{\text{scat}}$) are defined by normalizing the respective energy flows by the incident intensity. Cross section calculation thus involves evaluating a time-averaged Poynting vector, integrated over an appropriate surface. For $C_{\text{abs}}$ calculation, the total Poynting vector, including contributions from the incident field, is integrated, assuming the incident field crosses through $\Omega$. In contrast, to calculate $C_{\text{scat}}$, only the scattered field is considered; thus, the known incident field is excluded. Plane waves are not only incident fields treatable using the FDTD method. It is straightforward to implement arbitrary incident fields or include local radiation sources. For instance, including dipolar radiation sources enables calculations of how radiation from a localized emitter is altered by the presence of the metal nanoparticle. Moreover, computing other electromagnetic field-dependent physical quantities, such as optical forces, is also feasibly straightforward. Additionally, integrating more complex material responses, including optical nonlinearities, into simulations can be accomplished with relative ease.

FDTD technique represents a volumetric method that causes the demand for computational resources to surge in parallel with the enlargement of the system. This encompasses both the particles under study and areas where field calculations are required. Calculating fields can become computationally expensive, particularly when extrapolating to far-field characteristics. Moreover, systems encompassing multiple length scales present challenges when studied using uniform grids. The FDTD method also faces obstacles when dealing with media characterized by severe dielectric contrasts and high absorption rates typical of metallic materials. Despite these challenges, FDTD remains one of the most widely used computational techniques. This popularity can be attributed to its conceptual simplicity and the direct computation of electromagnetic fields. Additionally, it allows for the observation of temporal field evolutions through detailed animations and benefits from the availability of commercial software packages, along with the potential for efficient parallel processing on high-performance computing clusters.

### 4.2.2 *Discrete Dipole Approximation*

Real dielectric materials consist of polarizable atoms organized either in a regular crystalline structure or in an amorphous formation. The polarizability of the material stems from the interaction of the polarizabilities of each individual atom. Similarly, when individual nanoparticles are arranged in a particular configuration, their polarizations interact collectively, resulting in a comprehensive response for the entire setup. On the other hand, decomposing a complex nanostructure into simpler components allows for the overall response to be estimated based on the responses of these individual elements. This strategy underpins the discrete dipole approximation (DDA) method. In DDA, each structure to be analyzed is represented by a grid of discrete, interacting dipoles, where each dipole corresponds to the polarizability of the volume element of material it represents. Every dipole is influenced by external driving fields and the field effects from all other dipoles. In the frequency domain, the local field at any given dipole, denoted as dipole $i$, is calculated by considering both these internal and external contributions

$$\boldsymbol{E}_i^{\text{loc}}(\omega) = \boldsymbol{E}_i(\omega) + \sum_{j \neq i} \boldsymbol{G}_{ij}(\omega)\alpha_j(\omega)\boldsymbol{E}_j^{\text{loc}}(\omega) \tag{4.21}$$

where $\boldsymbol{E}_i$ represents the driving field at the location of dipole $i$, $\alpha_j$ is the polarizability of dipole $j$, and $\boldsymbol{G}_{ij}$ is the free-space Green's tensor that propagates the field from dipole $j$ to dipole $i$. To determine the structure's response, it is necessary to self-consistently compute the local field at each dipole. This is typically achieved by expressing the interactions as a matrix equation for all dipoles in the array and subsequently inverting the resultant large matrix. Once the local fields at each dipole are established, these values can be utilized to calculate the electric field at any point outside the array.

The DDA method, like the FDTD, is intuitively compelling and easily adaptable to various geometric structures. As a volume-based approach, DDA requires discretization of the structures themselves while the fields beyond these structures are calculated based on the emissions from the dipoles within. However, challenges arise when dealing with multiple length scales, particularly in achieving accurate field calculations near the surface of nanostructures, which demand high dipole densities. DDA bears conceptual resemblances to Green's function approaches, sharing similar benefits and drawbacks. In such methodologies, the material structures are considered as disturbances within the surrounding medium. The polarization effects caused by these disturbances generate fields, which are resolved using the Green's function to model free propagation in the media external to a point source.

### *4.2.3 Boundary Element Methods*

The computational demands of both the FDTD and DDA methods scale with the volume of the simulation domain. Conversely, the multiple multipole program (MMP) and boundary element method (BEM) scale with the surface area of interfaces within the system and can potentially offer superior computational efficiency. In both these approaches, the boundary conditions of Maxwell's equations at these interfaces are crucial to formulating a set of equations specific to each surface point. Solving these equations determines the spatial distribution of the field. Typically, these surface-related equations are resolved using a grid layout on the interfaces between different material zones. In both scenarios, the equations are predominantly solved within the frequency domain.

In the BEM, solving for the boundary conditions of an incident driving field involves utilizing effective charges and currents at each point on the surface grids. Similar to the DDA method, the fields at a specific surface point can be expressed in terms of the charges and currents at other surface points. This creates a large matrix equation that must be inverted to self-consistently determine these surface charges and currents. Once established, the fields throughout space are calculated using a Green's function approach to propagate the fields from the surface. Notably, the BEM simplifies the integration of intuitive physical concepts, particularly for understanding plasmon excitations, by determining effective surface charge densities. It accommodates variable grids efficiently, enabling areas with highly localized fields to be analyzed using a high-density grid while conserving computational resources in less critical regions. The BEM is especially effective for analyzing structures with smooth, easily definable surfaces. However, its application becomes more challenging with complex surfaces or those having sharp features and edges, where defining accurate boundary conditions and maintaining computational efficiency can be problematic.

### 4.2.4 Multiple Multipole Methods

In the MMP method, fields are characterized through multipolar expansions. This approach extends the conceptual framework of Mie theory commonly applied to spheres to accommodate more complex geometries. Typically, multiple multipolar expansions centered at various external points are employed to accurately define the field within a region, with each spatial segment described by distinct expansions. The coefficients for these expansions are ascertained by solving Maxwell's equations' boundary conditions at specific surface points that demarcate different regions. The effectiveness of this technique hinges on the clear definition of surfaces, akin to the requirements of the BEM method. For symmetrical structures, the locations and forms of multipolar expansions are readily determined, allowing a limited set of multipoles to succinctly describe the entire system. This provides a coherent and intuitive conceptualization of the field dynamics via the excited multipoles. Conversely, in structures with more complex geometries, determining the number, placement, and form of the necessary multipoles becomes inherently more challenging. Multiple expansion points can lead to an overcomplete representation, prompting numerous viable choices that render decisions arbitrary, complicate control, and diminish the physical intuition underlying the expansions.

To model heat generation and temperature distribution in thermoplasmonic systems, we can follow these steps: (1) Implement a physical model that defines the geometry of the nanoparticles and the surrounding medium, along with the material properties, such as dielectric functions dependent on frequency. (2) Use the MMP to calculate how incident electromagnetic waves, typically in the visible to near-infrared range, interact with the plasmonic nanoparticles. The resonant interactions correspond to surface plasmon excitations. (3) The plasmonic resonance leads to enhanced electromagnetic fields at specific frequencies. Part of this energy is dissipated as heat due to absorption in the metal. Calculate local heating by integrating power dissipation over the volume of the nanoparticles. (4) Couple the electromagnetic solution with thermal diffusion equations. We will need to solve the heat equation over time to find out how the generated heat diffuses into the surrounding medium. (5) After that, analyze the temperature distribution, hot spot locations, and temporal evolution of the heat profile.

## 4.3 Nanoscale Thermometry

Accurate detection of localized temperature changes is crucial in understanding thermoplasmonics [25]. An ideal nanothermometer should not only deliver precise measurements, but also cater to a wide temperature gradient and function reliably across varying environmental conditions. Moreover, its readout duration should be efficient enough to detect sudden changes in temperature. This necessitates nanothermometry possessing superior spatial, temporal, and thermal resolutions [26]. Significant progress has been made toward achieving this, and the reported techniques are principally divided into three categories: contact methods such as scanning microscopy or microthermocouple techniques, exemplified by scanning thermal microscopy (SThM); noncontact optical methods including IR thermography, Raman spectroscopy, thermoreflectance, interferometry, and so on; and semi-contact techniques using luminescent nanoparticles, such as organic dyes, proteins, quantum dots, nanodiamonds, and lanthanide-ion-doped nanoparticles [27, 28]. Each technique showcases particular merits and certain limitations. Therefore, choosing an optimal technique depends on the specific application. In this section, we categorize methods based on thermal radiation and Raman scattering to guide readers in selecting a suitable nanothermometry method. Furthermore, we discuss techniques according to whether they incorporate fluorescent molecular probes, nanoprobes, or temperature-induced shifts in the environment.

### 4.3.1 Thermal Radiation Spectroscopy

The spectrum of thermal (blackbody) radiation is determined solely by temperature. A particle temperature can be ascertained by equilibrating the absorbed and emitted power with heat dissipation. Planck's law allows for the calculation of emission intensity at a specific wavelength. However, conventional thermal imaging, commonly used for measuring the heating of nanoparticles in suspension, is unsuitable for nanothermometry [29, 30]. The issue arises from the peak intensity wavelength being greater than 10 $\mu$m, resulting in significantly reduced spatial resolution [31]. While light can propagate and be detected well away from its source, heat, on the other hand, diffuses. Consequently, heat from an NP is only detectable through near-field measurements. For this reason, initial efforts to probe temperature fields at microscopic levels utilized localized, small composite tips that function as nanoscale thermocouples.

In the 1990s, the so-called scanning thermal microscopy (SThM) was developed to collect both thermal images and topographical data. The layout of a SThM setup is depicted in Fig. 4.2 [32]. The core of SThM system is the thermal probe, which dictates the system's capabilities, operational range, and measurement accuracy. Typical sensing mechanisms in thermal probes include Seebeck thermovoltage, changes in electrical resistance, fluorescence, or thermal expansion [33]. The critical aspects of a SThM probe, whether based on thermoelectric or resistive technology, are the tip radius and the probe material. The tip radius not only affects the spatial resolution but also limits the heat transfer between the tip and the sample. The performance of the SThM measurements heavily depends on the material of the probe tip since it governs the thermal transduction mechanism. Furthermore, SThM probes can be made compatible with systems that measure optical properties, particularly when integrated with infrared technology.

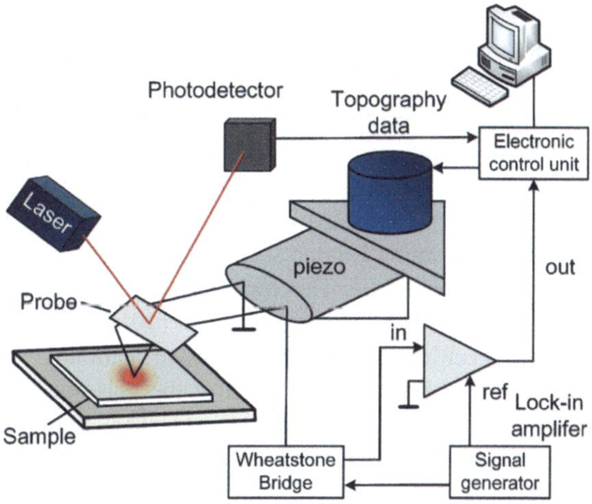


**Fig. 4.2** SThM setup, featuring a photodetector and a laser. These components are utilized to ascertain the position of the probe and the contact force by measuring the deflection of the cantilever, which is determined by the reflection of the laser. Adapted with permission from Ref. [32]. Copyright 2013 Elsevier

The SThM technique has been employed for thermal simulations and measurements on a silicon plasmonic waveguide excited in the infrared [34]. It is noteworthy that the authors did not extend their investigation to nanometric plasmonic particles, which could have demonstrated an enhanced spatial resolution with SThM. One plausible explanation for not exploring smaller systems might be the potentially invasive nature of SThM when applied to nanoobjects [35]. Despite offering high spatial resolution (~ 10 nm) and high sensitivity (~ 10 mK), SThM tends to be invasive and only allows for surface temperature detection [36]. Additionally, it necessitates slow scanning rates to generate thermal images and requires complex equipment and stringent measuring conditions, such as a high vacuum environment, which restrict its practical applications.

### 4.3.2 *Raman Spectroscopy*

Raman scattering involves the inelastic scattering of photons, where energy is exchanged between the photons and scatterers. This exchange corresponds to the vibrational states of the molecule involved in the scattering process. During Raman scattering, the energy of the scattered photon is defined by $h\nu' = h\nu - h\nu_m$, with $\nu$ representing the frequency of the incident light and $\nu_m$ indicating the frequency associated with the vibrational state of excitation. Conversely, there also exists a mechanism whereby $h\nu' = h\nu + h\nu_m$, known as anti-Stokes scattering. As such, in the spectral distribution of Raman scattering, the energy positions of the Stokes and anti-Stokes peaks appear symmetrically arranged around the energy of the excitation photon [37, 38]. Additionally, the intensity ratio of anti-Stokes ($I_{aS}$) to Stokes ($I_S$) peaks correlates with the occupation numbers of the ground and excited states, according to Fermi–Dirac statistics: [39]

$$\frac{I_{aS}}{I_S} = A\left(B + \exp\frac{-h\nu_m}{k_B T}\right) \tag{4.22}$$

In this context, $A$ represents an asymmetry parameter that accounts for the differing cross sections between Stokes and anti-Stokes shifts. $B$ is a parameter that describes vibrational pumping, which depends on both the intensity of the laser and the lifetime of the vibrational excited state. Since the second term is exclusively dependent on temperature, Raman scattering spectroscopy serves as a reliable method for measuring temperature. This technique has been effectively implemented in studies involving core–shell gold nanoparticles [40], gold nanorods [41], and nanodisks [42].

When exciting a gold nanorod with a monochromatic beam at its resonance frequency, a collective oscillation of the plasmons is induced. The decay of these plasmons results in the formation of electron–hole pairs, each with energy equal to that of the laser photon, hv. These hot electron–hole pairs have a low probability of recombining and may emit photoluminescence through either a Stokes or an anti-Stokes process. However, in both cases, the emitted energies cannot surpass $h\nu + k_{\rm B}$T. If the excitation coincides with the plasmon resonance, the resulting spectrum is expected to mimic the plasmon spectrum, scaled by a factor dictated by Bose–Einstein statistics due to the phonon population. It is posited that phonon coupling predominates over carrier–carrier interactions, as indicated by previous equations where occupation numbers conform to Fermi statistics. Given this, the emission intensity should correlate with the phonon occupation number n for anti-Stokes processes and $n + 1$ for Stokes processes. Consequently, the profile of the anti-Stokes emission spectrum can be derived and formatted according to the phonon distribution [43].

$$I(\nu') = I_{\rm SPR}(\nu')\left(\exp\frac{h\nu_m}{k_{\rm B} T} - 1\right)^{-1} \tag{4.23}$$

In this equation, $I(\nu')$ represents the emitted intensity, and $I_{\rm SPR}(\nu')$ denotes the surface plasmon resonance spectrum. The sole free parameters left to determine are the temperature $T$ and a normalization constant, which is not incorporated in this equation. Figure 4.3 illustrates a schematic of this method.

This model primarily hinges on only one free parameter: the absolute temperature of the particle. Consequently, this technique does not necessitate any calibration. By employing this method with an integration time of one minute, the authors achieved temperature accuracy of 6% [43]. Typically, the Stokes and anti-Stokes spectra are collected using a scanning confocal microscope attached to a spectrometer, with the excitation laser being filtered out by a notch filter (Fig. 4.3). This technique not only allows for monitoring the temperature of a nanoparticle during excitation but also enables the estimation of the temperature prior to excitation. Thus, it offers a valuable tool for probing local temperatures in vivo. The utility of this approach has further been validated in determining the temperature of gold bow tie nanoantennas [26, 44]. However, there is still a need for validation with less luminescent nanostructures and developing less invasive implementations before this technique can be routinely used for in vivo measurements.

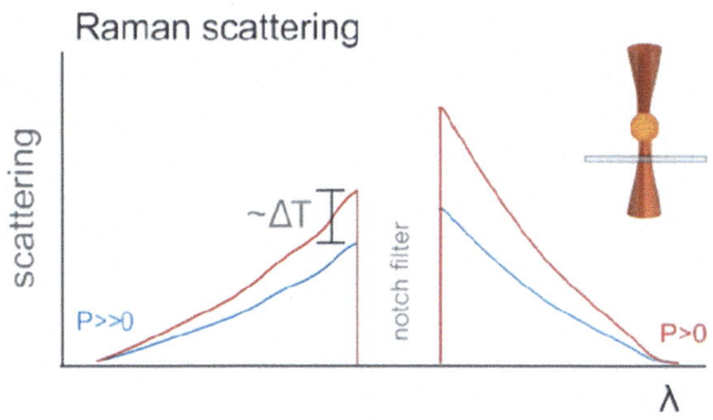


**Fig. 4.3** Nanothermometry employing Raman spectroscopy involves analyzing NP emission spectra at two distinct laser powers, $P$. The defining parameter is the intensity ratio between the anti-Stokes emissions (at lower wavelengths) and Stokes emissions (at higher wavelengths). From this ratio, one can deduce the temperature increase, $\Delta T$. Adapted from Ref. [43]. Copyright 2018 American Chemical Society

### 4.3.3 *Fluorescent Molecular Probes*

This technique uses fluorescent reporter molecules distributed throughout the medium surrounding a heated nanoparticle, as illustrated in Fig. 4.4a [45]. The fluorescence characteristics of these probe molecules, such as emission intensity, emission spectrum, or polarization anisotropy, are influenced by temperature. This influence is due to the temperature-dependent lifetime of the fluorescent state [46, 47] and the impact of Brownian dynamics on light emission [48].

*Fluorescence Polarization.* The temperature distribution surrounding plasmonic nanoparticles can be assessed using an optical method that relies on measuring the

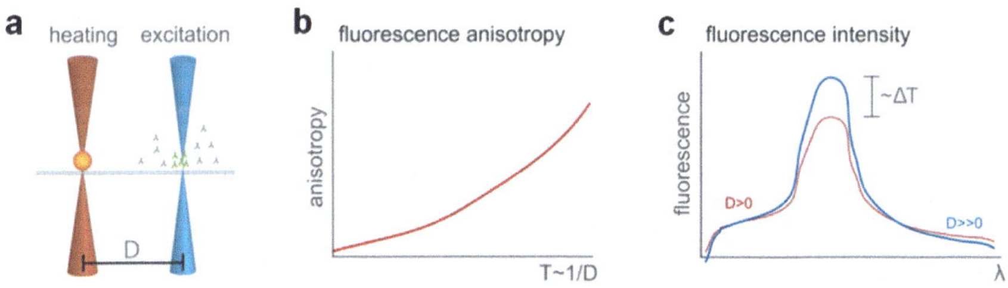


**Fig. 4.4** Nanothermometry using fluorescent molecular probes. **a** This diagram shows the setup, which includes a heating laser, a spherical nanoparticle, and a separate laser for exciting the fluorescent molecular probes. D denotes the distance between the focal points. **b** A graph depicting the fluorescence anisotropy of the fluorescent molecular probes as a function of distance. Figure adapted from Ref. [49] with permission from the Optical Society of America, copyright 2009. **c** A plot of fluorescence intensity versus wavelength $\lambda$, recorded at various distances from the heating laser. Adapted from Ref. [52]. Copyright 2016 Optical Society of America

molecular fluorescence polarization anisotropy [49, 50]. In this technique, fluorescent molecules are dispersed in the fluid around the NPs and are exposed to linearly polarized light. Since the orientation of each fluorescent molecule's dipole is random, the fluorescence emitted by a group of these illuminated molecules becomes partially depolarized. As the kinetic energy of the fluorescent molecules near the heated NPs rises, their rotational motion quickens, leading to further depolarization of the emitted fluorescence during their fluorescence lifetime as the molecules rotate more rapidly [51–53]. Consequently, thermal data are converted into optical signals via these fluorescent molecules (see Fig. 4.4b). For the effectiveness of this method, Brownian motions must be dampened, either by increasing the medium's viscosity or by enlarging the size of the proteins, aligning the timescale of molecular rotation with the fluorescence lifetime. This measurement strategy, which focuses on the ratio of light intensities with differing polarizations, remains unaffected by photobleaching. Thus, this technique has been introduced to gauge plasmonic heating in vivo in *C. elegans* [54, 55].

*Fluorescence Intensity.* A novel approach was developed utilizing nucleic acid oligonucleotides coated with fluorophores on gold nanoparticles to serve as a thermometer. This system operates based on DNA oligomers tagged at one end with a dedicated fluorophore and anchored at the other end to Au NPs using thiol-based chemistry. At ambient temperature, these complexes maintain a closed configuration, resulting in fluorescence quenching by the Au NP. Upon heating, the nucleic acid structure expands, disrupting the quenching effect. Consequently, the increase in fluorescence intensity is directly correlated with temperature [56]. Nevertheless, despite calibration by the researchers, this method still needs to validate its accuracy for precise nanothermometry or thermometric applications within living organisms. One limitation of this approach is its vulnerability to fluctuations in salt concentration and pH levels. Given the diverse nature of cellular environments, fine-tuning such a system could be challenging. Additionally, the temperature-sensitive fluorescence of tryptophan has been applied as an in-situ fluorescent thermometer for gauging photothermal conversion efficiency in bulk solutions containing Au NPs ranging from approximately 20–80 nm in diameter [57]. Although this fluorescence technique falls short in measuring individual particles, it can effectively measure optical heating across bulk nanoparticle groups with uniform optical densities.

A two-laser system has also been suggested for nanothermometry. This system employs a near-infrared laser to simultaneously trap and heat micron-sized gold shells while a second laser, at 445 nm, excites fluorescence from two dyes (Fig. 4.4c) [52]. In this setup, the point of fluorescence excitation by the second laser is spatially offset from the particle. The heated environment alters the emission ratios between a temperature-sensitive dye (rhodamine B) and a temperature-insensitive dye (rhodamine 110). A critical advantage of this method is its resilience to bleaching; the emission ratio exhibits a linear relationship with temperature changes, which can be calibrated by incrementally heating the sample in steps of 1 K. The authors recorded a significant heating rate of 360 K/W for the study particle. This example illuminates that strategies relying on temperature-dependent fluorescence typically require monitoring fluorophore intensity shifts in a solution. One challenge with this technique is that the irradiation of nanoparticles often produces insufficient heat for detecting discernible fluorescence changes using confocal microscopy. However, the notable scale of the gold shells employed in this experiment allows for detectable variations in intensity across the full focal volume, demonstrating the potential efficacy of this technique in specialized contexts like microscopic temperature measurements [52].

### *4.3.4 Nanoprobes*

Techniques that use reporter particles such as quantum dots, upconverting nanoparticles, or nanodiamonds provide a robust framework to measure heating at the single nanoparticle level. Leveraging a dual-laser system and illustrated in Fig. 4.5, these methods do not rely on fluorescence intensity. Instead, they are characterized by two key measurements: (i) spectral shifts and (ii) spectral line distortion. This approach ensures precise thermal monitoring by tracking changes in the optical properties of the reporter particles.

*Quantum Dots*: Quantum dots have been successfully employed as temperature nanoprobes in a setup involving laser-irradiated gold nanorods [58]. It has been observed that the fluorescence spectral lineshape of CdSe quantum dots exhibits a red shift when heated, with a rate of change measured at $0.1 \pm 0.05$ nm/K. Notably, this method's reliance on spectral shifts reduces its sensitivity to bleaching effects. The experimental design includes a flow system and a confocal microscope equipped with

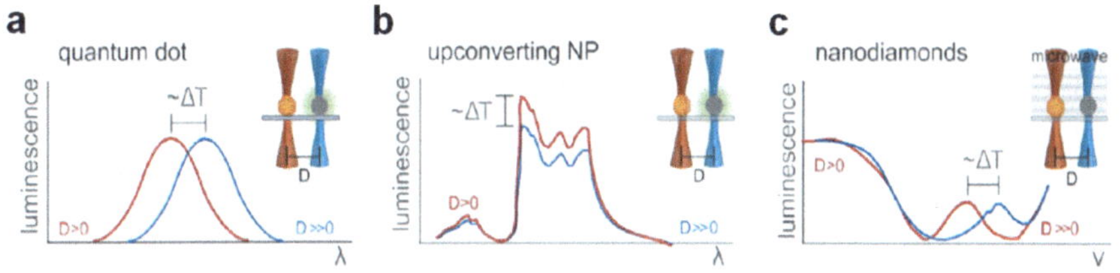


**Fig. 4.5** Nanothermometry using nanoprobes. **a** Quantum dot-based technique: This approach involves generating two-photon emission spectra from quantum dots (illustrated in gray), captured at two distinct distances, *D*, from a heating laser. The vertical dashed lines highlight the signals used to deduce the temperature increase, $\Delta T$. **b** Upconverting NP-based technique: this technique employs two-photon emission spectra from upconverting nanoparticles (shown in gray), measured at varying distances, from the heating laser. Parts A and B are adaptions from Ref. [62]. Copyright 2013 Elsevier Publishing. **c** Nanodiamond-based technique: optically detected spin resonance spectra from nanodiamonds (depicted in gray) are collected at different distances, from the heating laser. The vertical dashed lines represent the splitting parameter, which is crucial for determining the temperature increase $\Delta T$. Adaption from Ref. [66]. Copyright 2010 American Chemical Society. The inset displays the experimental setup, including a heating laser, a spherical nanoparticle, and a second laser utilized to excite the nanoprobes. *D* signifies the distance between the focal points of the two lasers—one for heating (red) and the other for excitation (blue)

dual lasers—one for plasmonic excitation of the gold nanorods and another for excitation of the quantum dots. This configuration allows for precise point temperature measurements within a microfluidic channel. Temperature increases ranging from 1 to 40 K were documented in the experiment. Further innovations were introduced by aligning both laser beams through separate microscope objectives on opposite sides of the microfluidic channel (Fig. 4.5a). This refined technique facilitated the exploration of the photothermal properties of a broader array of nanoparticles, including carbon nanotubes, nanocages, nanoshells, and nanostars [59, 60]. The primary objective in these studies was to conduct point measurements for investigating the photothermal characteristics of the nanostructured materials. While potentially compatible with a scanning system, a significant limitation of this spectrum-based approach is the time requirement for acquiring and processing an image, as each pixel's data necessitates individual spectrum measurement and analysis.

*Lanthanide Upconverting Nanoparticles*: Semiconductor nanoparticles or quantum dots are renowned for their ability to emit light of shorter wavelength than the excitation wavelength through a two-photon absorption process. This procedure relies on high photon fluxes and the simultaneous absorption of two photons. Conversely, lanthanide-doped nanocrystals function to convert infrared light into visible light through an anti-Stokes, multistep process. This behavior is opposite to the typical Stokes-shift luminescence, where an excitation photon with a shorter wavelength (and higher energy) results in the emission of a photon with a longer wavelength (and lower energy). Photon upconversion is facilitated by metastable states that enable sequential energy absorption, and materials that exhibit this property are broadly classified as upconverting nanoparticles (UCNPs). The distinguish features of UCNPs are their extensive lifetime, ranging from microseconds to milliseconds, which makes them well-suited for temperature-dependent spectroscopy applications (Fig. 4.5b) [61, 62]. Both the fluorescence intensity and lifetime, as well as the spectra of lanthanides, are influenced by temperature and the chemical environment, including parameters such as pH [63].

In 2013, researchers adopted a method using lanthanide-based compounds for temperature measurement in biological tissues. They used nanoparticles composed of a core of $Nd^{3+}$-doped $LaF_3$ (core) NPs encased by a nondoped $LaF_3$ shell. These were excited at 808 nm, emitting light at wavelengths of approximately 885 and 1060 nm [64]. The choice of these NPs was strategic, as their infrared absorption and emission spectra aligned well with the biological transparency window, allowing for accurate temperature measurements directly within biological tissues. The thermal sensitivity of this particular sensor reached approximately $\pm$ 2 K. Investigations into the brightness of the $Nd^{3+}$:$LaF_3$ probes revealed that they were significantly more luminous-over ten times brighter than a colloidal solution of CdTe quantum dots under comparable conditions. This enhanced efficiency was due to the fact that the $Nd^{3+}$:$LaF_3$ NPs utilize a one-photon excitation process, whereas the CdTe quantum dots require a two-photon excitation mechanism via virtual states [65]. Further comparisons made include other UCNPs, specifically those co-doped with

$Er^{3+}$ and $Yb^{3+}$ in $NaYF_4$ nanocrystals, set against CdTe quantum dots. These observations highlighted temperature resolutions of 1 K for UCNPs and 0.5 K for the quantum dots. The thermal sensitivity of UCNPs is driven by relative intensity shifts between two emission spectral peaks, whereas the quantum dots' sensitivity relates to spectral shifting [62]. This factor predisposes the lanthanide NP system to be more vulnerable to bleaching compared to quantum dots.

This technique was enhanced by integrating UCNPs composed of $NaGdF_4{:}Yb^{3+}20\%$, $Er^{3+}2\%$, with gold nanorods having an aspect ratio of 2:10, for advanced nanothermometry [61]. The localized surface plasmon resonance of the Au NRs (~ 980 nm) aligns with the absorption wavelength of $Yb^{3+}$ first excited state, resulting in substantial plasmonic resonant amplification. The AuNRs serve a dual purpose: they facilitate collective optical heating and enhance luminescence, with the latter effect varying based on the proximity between the Au NRs and the UCNPs. The upconversion interaction between $Yb^{3+}$ and $Er^{3+}$ ions yields distinct emission bands: a red emission at 654 nm and two green emissions at 526 and 533 nm. The temperature sensitivity of this method stems from the ratio of intensities between these two green emissions, referred to as S and H, respectively [67]. This relationship is utilized in a nanothermometry technique known as luminescence intensity ratio (LIR), where the intensities of these transitions, $I_H$ and $I_S$, correlate with Boltzmann's factor:

$$\frac{I_H}{I_S} \propto \exp\left(-\frac{\Delta E}{k_B T}\right) \tag{4.24}$$

here, $\Delta E$ represents the energy difference between the two emissive levels, $k_B$ is Boltzmann's constant, and $T$ denotes the temperature. The corresponding ratio, crucial for this analysis, was computed using data from the power-dependent measurements illustrated in Fig. 4.5b. Additionally, the longitudinal surface plasmon resonance of the gold nanorods is precisely aligned with the absorption wavelength of the UCNPs, enabling their simultaneous excitation. This synchronization allows the Au NRs to not only amplify the local field but also elevate the local temperature, which is calculated according to Eq. (4.24). Consequently, this alignment facilitates the detection of in-situ temperatures through optical means [61].

An alternative method involves optically trapping temperature-sensitive micron-sized particles within a fluid medium and utilizing them as probes to measure temperature fields in three dimensions. For instance, colloidal $NaYF_4{:}Er^{3+}$, $Yb^{3+}$ particles have been employed as micro-temperature probes within living cells heated by gold nanoparticles [68]. A 980 nm laser was used to trap the UCNPs and simultaneously induced a vibrant anti-Stokes emission in the visible spectrum, featuring two distinct narrow peaks at 525 and 550 nm. The intensity ratio of these peaks is highly dependent on temperature. Additionally, erbium oxide nanoparticles, sized at 150 nm, were also optically trapped and used as temperature probes. In this research, an image was produced where each pixel represented the emission spectrum of the optically trapped UCNP [69]. However, a potential issue with these methods is that the same laser is used for both trapping and heating.

*Nanodiamonds*: Nanodiamonds or hyperdiamonds are diamond nanoparticles that contain impurities or variations in their standard face-centered cubic crystal configuration. The unique fluorescence exhibited by nanodiamonds originates from these imperfections, referred to as color centers. Of particular note is the nitrogen vacancy (NV) color center, which occurs when a nitrogen atom and an adjacent vacancy replace two carbon atoms within the diamond's lattice structure. This specific color center emits a fluorescent glow in the red to near-infrared spectrum (550–800 nm) and exhibits exceptional photostability. While not as mainstream or widely utilized as quantum dots, nanodiamonds are increasingly being recognized for their potential, especially as highly efficient single-photon sources.

Fascinatingly, the spin states of a nitrogen vacancy (NV) center possess the unique ability to be optically polarized and precisely controlled using microwave radiation—a process recognized as optically detected magnetic resonance—and this remarkable capability extends even to the molecule level [26, 70]. When subjected to microwave exposure, NV centers exhibit a convenient optical transition in the visible spectrum between their triplet ground state, $|0\rangle$, and their excited states, $|\pm1\rangle$. Significantly, the intensity of fluorescence is closely linked to the microwave resonance. An increase in the triplet population results in a diminished fluorescence intensity. In the absence of a magnetic field, the $|\pm1\rangle$ states diverge from the $|0\rangle$ state due to a temperature-sensitive zero-field splitting [71]. A method for achieving precise spectroscopy involves applying an external oscillating electromagnetic field at a set frequency, followed by determining the detuning through the measurement of the probability of transitioning from the $|-1\rangle$ to the $|0\rangle$ states [72, 73].

In 2013, researchers applied this technique to nanoscale thermometry within a living cell [74]. The experiment involved spin coating a cover slip with gold nanorods (Au NRs) and single-crystalline nanodiamonds, each containing roughly 500 NV centers. When these Au NRs are heated by a resonant laser, the spin states of the NV centers can be efficiently manipulated using microwave pulses, and both initialized and detected through laser illumination. In the absence of an external magnetic field, the transition frequency between the $|0\rangle$ and $|\pm1\rangle$ states shows a temperature-dependent shift due to thermally induced lattice strains [71]. Consequently, the nanodiamonds were used to probe temperatures at varying distances from the Au NRs using electron spin resonance spectroscopy. This method successfully generated a detailed temperature map of the surface being examined.

Nanodiamonds are inert and considered nontoxic, making them suitable for use in nanothermometry within living cells. However, the traditional method of acquiring the entire spectrum for each pixel and then calculating the peak position is time-consuming. As a result, determining temperature profiles in systems far from equilibrium remains a challenging task. An innovative three-point sampling method has been developed for real-time temperature measurements of Au nanoparticles freely diffusing in water [75]. This approach leverages the observation that, in bulk diamonds, the widths of the optically detected magnetic resonance peaks remain almost constant, while the peak positions shift significantly-over 20 MHz-across temperatures ranging from 300 to 700 K [71, 74]. By assuming a constant width, it is possible to deduce temperature shifts by measuring changes at three predetermined frequencies in the fluorescence dip without having to scan the entire spectrum. This technique allows for the monitoring of temperature fluctuations over a range of ± 100 K in real time. Although the accuracy shifts slightly from 0.1 to 1 K, the duration required for measurement decreases from 30 s to a 10 microseconds.

This method was further refined to incorporate heat mapping along with a topographical scan of the sample. To capture spatial data, the nanodiamond needs to be maneuvered relative to the sample surface. This movement is typically facilitated using an atomic force microscope (AFM), where the nanodiamond is attached to the tip of the AFM. This combination is the sole technique that provides both detailed topography and quantum sensor-based measurements of the sample (Fig. 4.5c) [66]. In this setup, the authors placed Au nanoparticles on the sample surface. These nanoparticles emit a fluorescence signal that peaks when the nanodiamond is positioned directly above them, signaling significant absorption and consequent heating of laser light. To compile a temperature map, a complete spin resonance spectrum is recorded at each pixel of the scan while maintaining continuous laser excitation (250 mW) to ensure maximum heating. Temperature variations are then deduced by analyzing the spectroscopy data fits.

### *4.3.5 Surrounding Variation*

When a nanoparticle is heated, the heat dissipates into its surroundings, potentially altering their physical properties in detectable ways. For example, liquids change their refractive index when heated. Therefore, when a NP is locally heated in a solvent, this change in refractive index provides a method to optically detect local temperature variations. Furthermore, as discussed below, alterations in viscosity or temperature-induced phase transitions also manifest as detectable shifts in the surrounding environment, allowing for inferring the temperature of a plasmonic NP.

*Refractive Index Distortion*: By the temperature-induced change in refractive index, an optical microscopy technique for nanothermometry was developed using quadriwave lateral shearing interferometry [76]. The underlying principle is that an increase in temperature modifies the refractive index of a liquid, thereby distorting the planar wave front of incident light. This distortion can be quantitatively assessed using a wavefront analyzer (Fig. 4.6a). This method offers rapid data acquisition (~ 10 μs) and a sensitivity of 1 K [77], covering a broad temperature range beyond the limitations of traditional probe-based methods such as fluorescent molecules and quantum dots, which tend to degrade at temperatures nearing 100 °C [78]. The technique has been effectively applied to measure local steady-state temperature increases, heat source density, and the absorption cross section both in arrays of gold nanoparticles and complex nanowire systems. The plane wave passing through the area of interest experiences distortion due to the thermally induced refractive index changes in the liquid, which are then analyzed with an inversion algorithm to derive the temperature increase. Notably, by integrating experimental maps of heat power density, it is

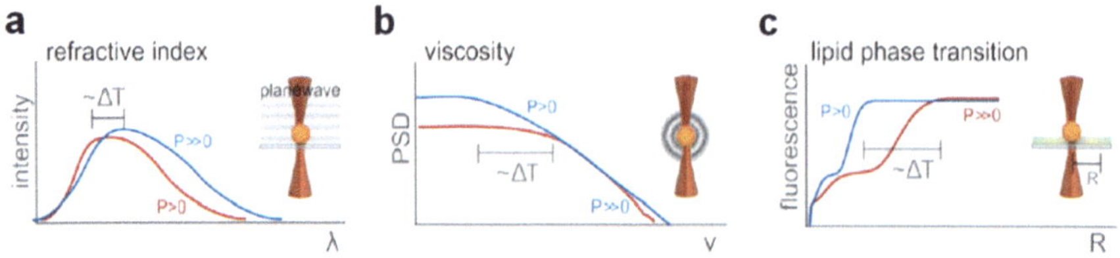


**Fig. 4.6** Nanothermometry leveraging changes in the surrounding environment: **a** Measurement of refractive index alterations at two different laser powers (labeled as P). Adapted with permission from Ref. [81]. **b** Assessment of viscosity changes through the analysis of power spectra density of an optically trapped nanoparticle upon two distinct laser powers. The key parameter used to deduce the change in temperature ($\Delta T$) is identified as the roll-off frequency. Adapted from Ref. [83]. Copyright 2006 Optical Society of America. **c** Detection of changes in lipid phase transitions determined by measuring light intensity as a function of distance ($R$) from the heated nanoparticle for two laser power settings. The area of the melted (fluorescent) region serves as the basis for inferring the temperature change. Adapted from Ref. [84]. Copyright 2010 American Chemical Society

possible to calculate the total power absorbed by the composite nanoparticle. Subsequently, the absorption cross section is determined by dividing the total absorbed energy by the incident irradiance, establishing that the steady-state temperature distribution directly correlates with the distance from the nanoparticle surface. This technique has also been extended to measure temperatures in thermoplasmonic nanowire systems and to determine the temperature threshold prior to bubble formation around heated plasmonic nanoparticles [79, 80]. It was found that temperatures can reach the superheated state of water (~ 220 °C), emphasizing the precision of this method in high-temperature applications.

The variation of refractive index was also applied to assess the temperatures of resonant gold nanoparticles [81]. Temperatures were ascertained by aligning the peak positions of experimentally acquired scattering spectra with those predicted by temperature-dependent scattering spectra, calculated using Mie theory. It was observed that the temperature increase is influenced by both the surrounding medium and the substrate. In 2015, an enhanced technique called plasmonic thermal microscopy was introduced, providing a more sensitive and rapid thermal sensing mechanism [82]. This method leverages the temperature dependency of the refractive index, detected through high-sensitivity surface plasmon resonance. Temperature measurements are made by correlating changes in temperature with shifts in the reflected angle of the plasmonic resonance of the heated nanoparticle. The primary advantage of this technique is its ability to discern temperature changes as subtle as 6 mK with a temporal resolution of 10 μs. However, this method is limited by it requires a gold film and thus cannot be universally employed to measure temperatures of nanostructures on more commonly used substrates like glass.

*Viscosity Changes*: One of the nanothermometry implementations involved using optically trapped silica and polystyrene particles, and it focused on analyzing the power dependence of their Brownian motions [85]. In this approach, the temperature rise was primarily attributed to the absorption of NIR light by the solution rather than the dielectric particle itself. Consequently, the temperature field in the fluid decayed radially around the particle. Spectral analysis of nanoparticle positional data was employed to determine the temperature within their own thermal gradient in the surrounding media [86]. The initial study on utilizing viscosity changes to gauge the temperatures of optically trapped gold nanoparticles appeared in 2006 (Fig. 4.6b) [83]. This work evaluated the stiffness of the optical trap using three distinct methods: (i) power spectral analysis, (ii) the equipartition theorem, and (iii) hydrodynamic drag calibration. Findings indicated that trap stiffness increased with higher laser powers. A significant discrepancy between trap stiffness values obtained from equipartition theorem calculations and power spectral analysis hinted at substantial heating effects. To estimate the temperature, the researchers postulated a relationship between temperature and viscosity, discovering temperature rises reaching up to 383 K at the particle's surface with a heating rate of 266 K/W for 100 nm Au NPs.

The diffusive behavior of nanoparticles within a thermal gradient, known as hot Brownian motion, has been further investigated in recent studies [87, 88]. This phenomenon can be employed to estimate the temperature of nanoparticles, exemplified by a study where the temperature of an optically trapped gold nanorod with an aspect ratio of 25/60 was determined [89, 90]. In this analysis, the Au NR oriented itself along the polarization direction of the trapping laser, leading to an angular dependency in the scattered light intensity. The ratio of spectral intensities can be approximated using

$$\frac{\langle I_{||} \rangle}{\langle I_{\perp} \rangle} \propto \frac{\kappa}{k_{\mathrm{B}} T} \tag{4.25}$$

In this context, $\langle I_{||} \rangle$ represents the light intensity scattered when oriented parallel, while $\langle I_{\perp} \rangle$ denotes the intensity from an orthogonal orientation. Here, $\kappa$ symbolizes the rotational stiffness of the trap, and $T$ indicates the effective temperature. Leveraging this relationship, the study identified a heating rate of 900 K/W for the optically trapped nanorod, aligning well with other documented instances of Au NR heating [91]. An alternative approach to deduce the heating of an Au NR involves analyzing both translational and rotational dynamics of a spinning Au NR optically confined in two dimensions, namely on a glass surface. By utilizing power spectral analysis along with luminescence decay time measurements, the authors estimated the temperature, finding heating rates that spanned from 1500 to 5000 K/W [92]. Therefore, additional research is necessary to refine this method for more general application.

*Lipid Phase Transitions*: The underlying principle of this technique involves positioning a nanoparticle on a lipid that has a clearly defined phase transition temperature and maintaining the system below this critical temperature. Subsequently, one should assess the size of the area that liquifies around the NP following laser exposure. An example of this methodology is depicted in Fig. 4.6c. As outlined in the original study from reference [84], the lipid bilayer was crafted from a saturated lipid with a phase transition temperature ($T_m$) of 33.8 °C. When NPs were exposed to radiation, the adjacent bilayer developed a circular fluid-phase zone, which could be detected by employing the fluorescent lipophilic tracer DiIC18, known to preferentially reside in the fluid phase. The radius of this melted sector, $D_m$, corresponds to the distance extending from the center of the NP to the boundary of the liquified region where the temperature equals $T_m$. To calculate the increase in temperature at the NP surface, Eq. 2.35 was utilized yielding $\Delta T(R) = \Delta T(D_m) \cdot D_m/R$. Utilizing this approach, heating rates of 452 and 1640 K/W were determined for Au NPs sized 100 and 200 nm, respectively. These values are notably higher than those previously reported [83], possibly due to the employment of the dipole approximation and the presupposed correlation between two variables, temperature and viscosity. This lipid bilayer melting technique has been further refined to evaluate how the heating of Au NRs varies depending on their alignment relative to the laser's polarization vector, as well as to analyze the heating dynamics of intricately designed composite surface-fabricated nanostructures [91, 93], where experimental findings align closely with theoretical projections.

Several techniques have been developed to measure the heating effects of nanoparticles moving or held in 3D environments, as noted in references [83, 94]. One prominent technique involved using the thermally triggered release from a giant unilamellar vesicle (GUV) loaded with fluorophores for the nanothermometry of optically trapped Au NPs (ranging from 80 to 200 nm) within a 3D setup [95]. In this assay, Au NPs were confined at different proximities to a GUV, and at a certain distance, $D_m$, the emitted heat from the captured AuNP increased the surrounding temperature of the GUV to $T_m$. This increase prompted a detectable efflux of molecules from the GUV. By analyzing $D_m$ and $T_m$, the complete temperature profile could be inferred, analogous to the methodology described in reference [84]. The precision of this assay is contingent on the sharpness of the phase transition temperature and the breadth of that transition. These parameters, the phase transition temperature and its range, can be determined through calorimetry and visualization using various lipophilic dyes [96]. A significant benefit of this method is its ability to directly determine temperatures without requiring knowledge about variables such as medium viscosity, thermal conductivities, or exact laser intensities. Moreover, the measurements can be conducted remotely from the irradiated NP, applicable in both 2D and 3D settings. However, a limitation of this approach is its compatibility. Not all nanoscopic systems may interact favorably with lipid bilayers. Additionally, the selected lipid bilayer may be susceptible to changes in osmotic pressure and the conditions of the surrounding solution.

To assist researchers in selecting the suitable nanothermometry for specific applications, it is important to note that the spatial resolution of many techniques reviewed is constrained by the diffraction limit of visible light, which is approximately 200–300 nm (Fig. 4.7) [27]. Various studies reporting temperature variations within living cells have cast doubt on the precision of intracellular thermometry methods [97–99]. This controversy underscores the necessity for rigorous calibration of nanothermometry and highlights the challenges posed by complex biological environments. A current trend in this field involves the development of methodologies that not only

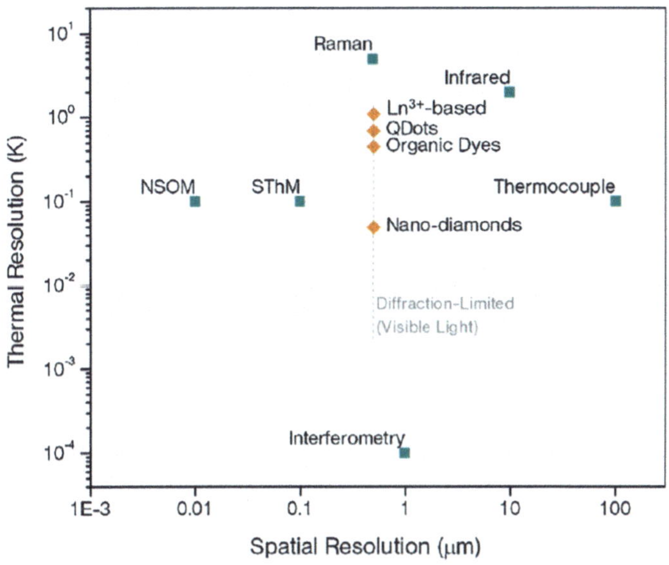


**Fig. 4.7** Comparative analysis of thermal versus spatial resolution across different thermometry techniques. Reproduced with permission from Ref. [27]. Copyright 2018 Elsevier

allow measurement of high temperatures but also offer extremely high temporal resolution [26]. Innovations like the combination of time-domain thermoreflectance and optical thermoreflectance imaging have recently advanced our ability to investigate the optical and thermal dynamics of plasmonic nanostructures with nanometer and nanosecond spatiotemporal resolution [100]. Despite these advancements, there remains significant potential for developing and refining methods capable of accurately measuring temperatures associated with or influenced by plasmonic nanostructures both in vitro and in vivo. These methods need to be precisely calibrated and tailored to address specific experimental conditions.

## References

1. Z.B. Wang, B.S. Luk'yanchuk, M.H. Hong, Y. Lin, T.C. Chong, Energy flow around a small particle investigated by classical Mie theory. Phys. Rev. B **70** (2004)
2. R. Yu, L.M. Liz-Marzan, F.J. Garcia de Abajo, Universal analytical modeling of plasmonic nanoparticles. Chem. Soc. Rev. **46**, 6710–6724 (2017)
3. M.C. Wang, E. Guth, On the theory of multiple scattering, particularly of charged particles. Phys. Rev. **84**, 1092–1111 (1951)
4. C. Sauvan, J.P. Hugonin, I.S. Maksymov, P. Lalanne, Theory of the spontaneous optical emission of nanosize photonic and plasmon resonators. Phys. Rev. Lett. **110**, 237401 (2013)
5. F. Wang, Y.R. Shen, General properties of local plasmons in metal nanostructures. Phys. Rev. Lett. **97**, 206806 (2006)

6. F.J.G. de Abajo, Nonlocal effects in the plasmons of strongly interacting nanoparticles, dimers, and waveguides. J. Phys. Chem. C **112**, 17983–17987 (2008)
7. F.J. García de Abajo, Graphene plasmonics: challenges and opportunities. ACS Photon. **1**, 135–152 (2014)
8. T.J. Davis, D.E. Gómez, Colloquium: an algebraic model of localized surface plasmons and their interactions. Rev. Modern Phys. **89** (2017)
9. P.T. Kristensen, S. Hughes, Modes and mode volumes of leaky optical cavities and plasmonic nanoresonators. ACS Photon. **1**, 2–10 (2013)
10. Q. Bai, M. Perrin, C. Sauvan, J.P. Hugonin, P. Lalanne, Efficient and intuitive method for the analysis of light scattering by a resonant nanostructure. Opt. Expr. **21**, 27371–27382 (2013)
11. D.M. Yu, Y.N. Liu, F.L. Tian, X.M. Pan, X.Q. Sheng, Accurate thermoplasmonic simulation of metallic nanoparticles. J. Quant. Spectrosc. Radiat. Transf. **187**, 150–160 (2017)
12. N.A. Mortensen, Mesoscopic electrodynamics at metal surfaces. Nanophotonics **10**, 2563–2616 (2021)
13. J.B. Schneider, Understanding the finite-difference time-domain method, School of electrical engineering and computer science. Washington State University (2010)
14. F.L. Teixeira, C. Sarris, Y. Zhang, D.Y. Na, J.P. Berenger, Y. Su, M. Okoniewski, W.C. Chew, V. Backman, J.J. Simpson, Finite-difference time-domain methods. Nat. Rev. Methods Prim. **3** (2023)
15. M.A. Yurkin, A.G. Hoekstra, The discrete-dipole-approximation code ADDA: capabilities and known limitations. J. Quant. Spectrosc. Radiat. Transfer **112**, 2234–2247 (2011)
16. B.T. Draine, P.J. Flatau, Discrete-dipole approximation for scattering calculations. J. Opt. Soc. Am. A **11**, 1491 (1994)
17. S. Rotter, S. Gigan, Light fields in complex media: mesoscopic scattering meets wave control. Rev. Modern Phys. **89** (2017)
18. G. Baffou, R. Quidant, C. Girard, Thermoplasmonics modeling: a Green's function approach, Phys. Rev. B **82** (2010)
19. C. Hafner, R. Ballisti, The multiple multipole method (Mmp). COMPEL: Int. J. Comput. Math. Electr. Electron. Eng. **2**, 1–7 (1983)
20. N.G. Khlebtsov, T-matrix method in plasmonics: an overview. J. Quant. Spectrosc. Radiat. Transfer **123**, 184–217 (2013)
21. K.M. Leung, Y.F. Liu, Photon band structures: the plane-wave method. Phys. Rev. B Condens. Matter **41**, 10188–10190 (1990)
22. A. Amirjani, P.Z. Abyaneh, P.A. Masouleh, S.K. Sadrnezhaad, Finite and boundary element methods for simulating optical properties of plasmonic nanostructures. Plasmonics **17**, 1095–1106 (2022)
23. G. Baffou, R. Quidant, F.J. Garcia de Abajo, Nanoscale control of optical heating in complex plasmonic systems. ACS Nano **4**, 709–716 (2010)
24. G.B. Matthew Pelton, *Introduction to Metal-Nanoparticle Plasmonics* (Wiley, 2013), p. 275
25. G. Baffou, Anti-stokes thermometry in nanoplasmonics. ACS Nano **15**, 5785–5792 (2021)
26. L. Jauffred, A. Samadi, H. Klingberg, P.M. Bendix, L.B. Oddershede, Plasmonic heating of nanostructures. Chem. Rev. **119**, 8087–8130 (2019)
27. M. Quintanilla, L.M. Liz-Marzán, Guiding rules for selecting a nanothermometer. Nano Today **19**, 126–145 (2018)
28. H. Zhou, M. Sharma, O. Berezin, D. Zuckerman, M.Y. Berezin, Nanothermometry: from microscopy to thermal treatments. Chemphyschem : a Eur. J. Chem. Phys. Phys. Chem. **17**, 27–36 (2016)
29. S. Jones, D. Andrén, P. Karpinski, M. Käll, Photothermal heating of plasmonic nanoantennas: influence on trapped particle dynamics and colloid distribution. ACS Photon. **5**, 2878–2887 (2018)
30. V.P. Pattani, J.W. Tunnell, Nanoparticle-mediated photothermal therapy: a comparative study of heating for different particle types. Lasers Surg. Med. **44**, 675–684 (2012)
31. L. Jauffred, Why not use thermal radiation for nanothermometry? Appl. Opt. **57**, 9508–9511 (2018)

32. R. Heiderhoff, H. Li, T. Riedl, Dynamic near-field scanning thermal microscopy on thin films. Microelectron. Reliab. **53**, 1413–1417 (2013)
33. Y. Zhang, W. Zhu, F. Hui, M. Lanza, T. Borca-Tasciuc, M. Muñoz Rojo, A review on principles and applications of scanning thermal microscopy (SThM). Adv. Funct. Mater. **30** (2019)
34. B. Desiatov, I. Goykhman, U. Levy, Direct temperature mapping of nanoscale plasmonic devices. Nano Lett. **14**, 648–652 (2014)
35. K. Nam, H. Kim, W. Park, J.S. Ahn, S. Choi, Probing the optical near-field of plasmonic nano structure using scanning thermal microscopy. Nanotechnology **34** (2022)
36. M.L. Debasu, D. Ananias, I. Pastoriza-Santos, L.M. Liz-Marzan, J. Rocha, L.D. Carlos, All-in-one optical heater-thermometer nanoplatform operative from 300 to 2000 k based on $Er^{3+}$ emission and blackbody radiation. Adv. Mater. **25** (2013) 4868-4874
37. X. Wang, S.C. Huang, S. Hu, S. Yan, B. Ren, Fundamental understanding and applications of plasmon-enhanced Raman spectroscopy. Nat. Rev. Phys. **2**, 253–271 (2020)
38. S.Y. Ding, J. Yi, J.F. Li, B. Ren, D.Y. Wu, R. Panneerselvam, Z.Q. Tian, Nanostructure-based plasmon-enhanced Raman spectroscopy for surface analysis of materials. Nat. Rev. Mater. **1** (2016)
39. R.C. Maher, P.G. Etchegoin, E.C. Le Ru, L.F. Cohen, A conclusive demonstration of vibrational pumping under surface enhanced Raman scattering conditions. J. Phys. Chem. B **110**, 11757–11760 (2006)
40. A. Yashchenok, A. Masic, D. Gorin, O. Inozemtseva, B.S. Shim, N. Kotov, A. Skirtach, H. Mohwald, Optical heating and temperature determination of core-shell gold nanoparticles and single-walled carbon nanotube microparticles. Small **11**, 1320–1327 (2015)
41. J. Huang, W. Wang, C.J. Murphy, D.G. Cahill, Resonant secondary light emission from plasmonic Au nanostructures at high electron temperatures created by pulsed-laser excitation. Proc. Natl. Acad. Sci. U.S.A. **111**, 906–911 (2014)
42. X. Xie, D.G. Cahill, Thermometry of plasmonic nanostructures by anti-Stokes electronic Raman scattering. Appl. Phys. Lett. **109** (2016)
43. A. Carattino, M. Caldarola, M. Orrit, Gold nanoparticles as absolute nanothermometers. Nano Lett. **18**, 874–880 (2018)
44. S. Dodson, M. Haggui, R. Bachelot, J. Plain, S. Li, Q. Xiong, Optimizing electromagnetic hotspots in plasmonic bowtie nanoantennae. J. Phys. Chem. Lett. **4**, 496–501 (2013)
45. J. Zhou, B. Del Rosal, D. Jaque, S. Uchiyama, D. Jin, Advances and challenges for fluorescence nanothermometry. Nat. Methods **17**, 967–980 (2020)
46. S. Freddi, L. Sironi, R. D'Antuono, D. Morone, A. Dona, E. Cabrini, L. D'Alfonso, M. Collini, P. Pallavicini, G. Baldi, D. Maggioni, G. Chirico, A molecular thermometer for nanoparticles for optical hyperthermia. Nano Lett. **13**, 2004–2010 (2013)
47. S. Hormeno, P. Gregorio-Godoy, J. Perez-Juste, L.M. Liz-Marzan, B.H. Juarez, J.R. Arias-Gonzalez, Laser heating tunability by off-resonant irradiation of gold nanoparticles. Small **10**, 376–384 (2014)
48. S. Ito, T. Sugiyama, N. Toitani, G. Katayama, H. Miyasaka, Application of fluorescence correlation spectroscopy to the measurement of local temperature in solutions under optical trapping condition. J. Phys. Chem. B **111**, 2365–2371 (2007)
49. G. Baffou, M.P. Kreuzer, F. Kulzer, R. Quidant, Temperature mapping near plasmonic nanostructures using fluorescence polarization anisotropy. Opt. Express **17**, 3291–3298 (2009)
50. D. Semeniak, D.F. Cruz, A. Chilkoti, M.H. Mikkelsen, Plasmonic fluorescence enhancement in diagnostics for clinical tests at point-of-care: a review of recent technologies. Adv. Mater. **35**, e2107986 (2023)
51. G. Baffou, C. Girard, R. Quidant, Mapping heat origin in plasmonic structures. Phys. Rev. Lett. **104**, 136805 (2010)
52. M. Siler, J. Jezek, P. Jakl, Z. Pilat, P. Zemanek, Direct measurement of the temperature profile close to an optically trapped absorbing particle. Opt. Lett. **41**, 870–873 (2016)
53. A. Sharma, P. Verwilst, M. Li, D. Ma, N. Singh, J. Yoo, Y. Kim, Y. Yang, J.H. Zhu, H. Huang, X.L. Hu, X.P. He, L. Zeng, T.D. James, X. Peng, J.L. Sessler, J.S. Kim, Theranostic fluorescent probes. Chem. Rev. **124**, 2699–2804 (2024)

54. J.S. Donner, S.A. Thompson, M.P. Kreuzer, G. Baffou, R. Quidant, Mapping intracellular temperature using green fluorescent protein. Nano Lett. **12**, 2107–2111 (2012)
55. J.S. Donner, S.A. Thompson, C. Alonso-Ortega, J. Morales, L.G. Rico, S.I.C.O. Santos, R. Quidant, Imaging of plasmonic heating in a living organism. ACS Nano **7**, 8666–8672 (2013)
56. S. Ebrahimi, Y. Akhlaghi, M. Kompany-Zareh, A. Rinnan, Nucleic acid based fluorescent nanothermometers. ACS Nano **8**, 10372–10382 (2014)
57. M.J. Chiu, L.K. Chu, Quantifying the photothermal efficiency of gold nanoparticles using tryptophan as an in situ fluorescent thermometer. Phys. Chem. Chem. Phys. **17**, 17090–17100 (2015)
58. L.M. Maestro, P. Haro-González, J.G. Coello, D. Jaque, Absorption efficiency of gold nanorods determined by quantum dot fluorescence thermometry. Appl. Phys. Lett. **100** (2012)
59. L.M. Maestro, P. Haro-Gonzalez, A. Sanchez-Iglesias, L.M. Liz-Marzan, J. Garcia Sole, D. Jaque, Quantum dot thermometry evaluation of geometry dependent heating efficiency in gold nanoparticles. Langmuir: ACS J. Surf. Colloids **30**, 1650–1658 (2014)
60. L.M. Maestro, P. Haro-Gonzalez, B. del Rosal, J. Ramiro, A.J. Caamano, E. Carrasco, A. Juarranz, F. Sanz-Rodriguez, J.G. Sole, D. Jaque, Heating efficiency of multi-walled carbon nanotubes in the first and second biological windows. Nanoscale **5**, 7882–7889 (2013)
61. S. Rohani, M. Quintanilla, S. Tuccio, F. De Angelis, E. Cantelar, A.O. Govorov, L. Razzari, F. Vetrone, Enhanced luminescence, collective heating, and nanothermometry in an ensemble system composed of lanthanide-doped upconverting nanoparticles and gold nanorods. Adv. Opt. Mater. **3**, 1606–1613 (2015)
62. D. Jaque, L.M. Maestro, E. Escudero, E.M. Rodríguez, J.A. Capobianco, F. Vetrone, A. Juarranz de la Fuente, F. Sanz-Rodríguez, M.C. Iglesias-de la Cruz, C. Jacinto, U. Rocha, J. García Solé, Fluorescent nano-particles for multi-photon thermal sensing. J. Lumines. **133**, 249–253 (2013)
63. S. Premcheska, M. Lederer, A.M. Kaczmarek, The importance, status, and perspectives of hybrid lanthanide-doped upconversion nanothermometers for theranostics. Chem. Commun. **58**, 4288–4307 (2022)
64. U. Rocha, C. Jacinto da Silva, W. Ferreira Silva, I. Guedes, A. Benayas, L. Martinez Maestro, M. Acosta Elias, E. Bovero, F.C. van Veggel, J.A. Garcia Sole, D. Jaque, Subtissue thermal sensing based on neodymium-doped LaF(3) nanoparticles. ACS Nano **7**, 1188–1199 (2013)
65. L. Jauffred, L.B. Oddershede, Two-photon quantum dot excitation during optical trapping. Nano Lett. **10**, 1927–1930 (2010)
66. J.P. Tetienne, A. Lombard, D.A. Simpson, C. Ritchie, J. Lu, P. Mulvaney, L.C. Hollenberg, Scanning nanospin ensemble microscope for nanoscale magnetic and thermal imaging. Nano Lett. **16**, 326–333 (2016)
67. D. Wawrzynczyk, A. Bednarkiewicz, M. Nyk, W. Strek, M. Samoc, Neodymium(III) doped fluoride nanoparticles as non-contact optical temperature sensors. Nanoscale **4**, 6959–6961 (2012)
68. P. Rodriguez-Sevilla, Y. Zhang, P. Haro-Gonzalez, F. Sanz-Rodriguez, F. Jaque, J.G. Sole, X. Liu, D. Jaque, Thermal scanning at the cellular level by an optically trapped upconverting fluorescent particle. Adv. Mater. **28**, 2421–2426 (2016)
69. S. Baral, S.C. Johnson, A.A. Alaulamie, H.H. Richardson, Nanothermometry using optically trapped erbium oxide nanoparticle. Appl. Phys. A **122** (2016)
70. Z. Qin, Z. Wang, F. Kong, J. Su, Z. Huang, P. Zhao, S. Chen, Q. Zhang, F. Shi, J. Du, In situ electron paramagnetic resonance spectroscopy using single nanodiamond sensors. Nat. Commun. **14**, 6278 (2023)
71. V.M. Acosta, E. Bauch, M.P. Ledbetter, A. Waxman, L.S. Bouchard, D. Budker, Temperature dependence of the nitrogen-vacancy magnetic resonance in diamond. Phys. Rev. Lett. **104** (2010)
72. F. Dolde, H. Fedder, M.W. Doherty, T. Nöbauer, F. Rempp, G. Balasubramanian, T. Wolf, F. Reinhard, L.C.L. Hollenberg, F. Jelezko, J. Wrachtrup, Electric-field sensing using single diamond spins. Nat. Phys. **7**, 459–463 (2011)

73. P. Neumann, I. Jakobi, F. Dolde, C. Burk, R. Reuter, G. Waldherr, J. Honert, T. Wolf, A. Brunner, J.H. Shim, High-precision nanoscale temperature sensing using single defects in diamond. Nano Lett. **13**, 2738–2742 (2013)
74. G. Kucsko, P.C. Maurer, N.Y. Yao, M. Kubo, H.J. Noh, P.K. Lo, H. Park, M.D. Lukin, Nanometre-scale thermometry in a living cell. Nature **500**, 54–58 (2013)
75. Y.K. Tzeng, P.C. Tsai, H.Y. Liu, O.Y. Chen, H. Hsu, F.G. Yee, M.S. Chang, H.C. Chang, Time-resolved luminescence nanothermometry with nitrogen-vacancy centers in nanodiamonds. Nano Lett. **15**, 3945–3952 (2015)
76. G. Baffou, *Thermoplasmonics Heating Metal Nanoparticles Using Light* (Cambridge University Press, 2018). ISBN 978-971-108-41832-41834
77. G. Baffou, P. Bon, J. Savatier, J. Polleux, M. Zhu, M. Merlin, H. Rigneault, S. Monneret, Thermal imaging of nanostructures by quantitative optical phase analysis. ACS Nano **6**, 2452–2458 (2012)
78. P. Berto, E.B. Ureña, P. Bon, R. Quidant, H. Rigneault, G. Baffou, Quantitative absorption spectroscopy of nano-objects. Phys. Rev. B **86** (2012)
79. A.P. Bell, J.A. Fairfield, E.K. McCarthy, S. Mills, J.J. Boland, G. Baffou, D. McCloskey, Quantitative study of the photothermal properties of metallic nanowire networks. ACS Nano **9**, 5551–5558 (2015)
80. G. Baffou, J. Polleux, H. Rigneault, S. Monneret, Super-heating and micro-bubble generation around plasmonic nanoparticles under cw illumination. J. Phys. Chem. C **118**, 4890–4898 (2014)
81. K. Setoura, Y. Okada, D. Werner, S. Hashimoto, Observation of nanoscale cooling effects by substrates and the surrounding media for single gold nanoparticles under CW-laser illumination. ACS Nano **7**, 7874–7885 (2013)
82. Z.X. Chen, X.N. Shan, Y. Guan, S.P. Wang, J.J. Zhu, N.J. Tao, Imaging local heating and thermal diffusion of nanomaterials with plasmonic thermal microscopy. ACS Nano **9**, 11574–11581 (2015)
83. Y. Seol, A.E. Carpenter, T.T. Perkins, Gold nanoparticles: enhanced optical trapping and sensitivity coupled with significant heating. Opt. Lett. **31**, 2429–2431 (2006)
84. P.M. Bendix, S. Nader, S. Reihani, L.B. Oddershede, Direct measurements of heating by electromagnetically trapped gold nanoparticles on supported lipid bilayers. ACS Nano **4**, 2256–2262 (2010)
85. E.J. Peterman, F. Gittes, C.F. Schmidt, Laser-induced heating in optical traps. Biophys. J. **84**, 1308–1316 (2003)
86. P. Haro-Gonzalez, W.T. Ramsay, L. Martinez Maestro, B. del Rosal, K. Santacruz-Gomez, C. Iglesias-de la Cruz Mdel, F. Sanz-Rodriguez, J.Y. Chooi, P. Rodriguez Sevilla, M. Bettinelli, D. Choudhury, A.K. Kar, J.G. Sole, D. Jaque, L. Paterson, Quantum dot-based thermal spectroscopy and imaging of optically trapped microspheres and single cells. Small **9**, 2162–2170 (2013)
87. D. Rings, R. Schachoff, M. Selmke, F. Cichos, K. Kroy, Hot brownian motion. Phys. Rev. Lett. **105**, 090604 (2010)
88. J.Y. Mo, M.G. Raizen, Highly resolved brownian motion in space and in time. Annu. Rev. Fluid Mech. **51**, 403–428 (2019)
89. E. Ortiz-Rivero, S. Orozco-Barrera, H. Chatterjee, C.D. Gonzalez-Gomez, C. Caro, M.L. Garcia-Martin, P.H. Gonzalez, R.A. Rica, F. Gamez, Light-to-heat conversion of optically trapped hot Brownian particles. ACS Nano **17**, 24961–24971 (2023)
90. P.V. Ruijgrok, N.R. Verhart, P. Zijlstra, A.L. Tchebotareva, M. Orrit, Brownian fluctuations and heating of an optically aligned gold nanorod. Phys. Rev. Lett. **107**, 037401 (2011)
91. H. Ma, P.M. Bendix, L.B. Oddershede, Large-scale orientation dependent heating from a single irradiated gold nanorod. Nano Lett. **12**, 3954–3960 (2012)
92. F. Hajizadeh, L. Shao, D. Andrén, P. Johansson, H. Rubinsztein-Dunlop, M. Käll, Brownian fluctuations of an optically rotated nanorod. Optica **4**, 746 (2017)
93. H. Ma, P. Tian, J. Pello, P.M. Bendix, L.B. Oddershede, Heat generation by irradiated complex composite nanostructures. Nano Lett. **14**, 612–619 (2014)

94. A. Ohlinger, S. Nedev, A.A. Lutich, J. Feldmann, Optothermal escape of plasmonically coupled silver nanoparticles from a three-dimensional optical trap. Nano Lett. **11**, 1770–1774 (2011)
95. A. Kyrsting, P.M. Bendix, D.G. Stamou, L.B. Oddershede, Heat profiling of three-dimensionally optically trapped gold nanoparticles using vesicle cargo release. Nano Lett. **11**, 888–892 (2011)
96. T. Baumgart, G. Hunt, E.R. Farkas, W.W. Webb, G.W. Feigenson, Fluorescence probe partitioning between Lo/Ld phases in lipid membranes. Biochem. Biophys. Acta. **1768**, 2182–2194 (2007)
97. G. Baffou, H. Rigneault, D. Marguet, L. Jullien, A critique of methods for temperature imaging in single cells. Nat. Methods **11**, 899–901 (2014)
98. K. Okabe, N. Inada, C. Gota, Y. Harada, T. Funatsu, S. Uchiyama, Intracellular temperature mapping with a fluorescent polymeric thermometer and fluorescence lifetime imaging microscopy. Nat. Commun. **3**, 705 (2012)
99. T. Bai, N. Gu, Micro/Nanoscale Thermometry for Cellular Thermal Sensing. Small **12**, 4590–4610 (2016)
100. D. Wang, Y.R. Koh, Z.A. Kudyshev, K. Maize, A.V. Kildishev, A. Boltasseva, V.M. Shalaev, A. Shakouri, Spatial and temporal nanoscale plasmonic heating quantified by thermoreflectance. Nano Lett. **19**, 3796–3803 (2019)

# Chapter 5
# Engineering Applications

This section discusses the diverse applications of thermoplasmonics including life sciences, optoelectronic information science, energy, chemistry, environmental studies, and material science. Specific areas of application include protein denaturation, photothermal cancer therapy, drug and gene delivery, heat-assisted magnetic recording, photoacoustic imaging, plasmonic-induced nanochemistry, photothermal imaging, solar steam generation, and experiments with single living cells. Each application is discussed in detail, focusing on the origins of the associated techniques, key pioneering studies that introduced significant variations or new concepts, the current advancements in the field, and the challenges that persist.

## 5.1 Life Science

### *5.1.1 Photothermal Therapy*

Hyperthermia therapy leverages heat for treating diseases or pathologies. The biological and physiological responses vary based on the level of applied heat: below 41 °C, it stimulates the immune system and enhances cellular membrane diffusion; between 41 and 43 °C, it causes partial cellular damage and increases blood flow; between 43 and 50 °C, it leads to irreversible cellular damage and protein unfolding; and above 50 °C, it results in protein denaturation and DNA damage [1]. In this domain, thermoplasmonics significantly enhances light-to-heat conversion in the near-infrared spectrum, where tissue transparency is higher, presenting itself as a highly effective therapeutic approach [2]. In photothermal therapy (PTT), activation of plasmonic agents through external laser irradiation at precise wavelengths allows heat to build up specifically in lesions, accurately targeting tumors (Fig. 5.1a) [3]. When selecting the optimal plasmonic particles, several factors need consideration: efficient light-to-heat conversion in the NIR spectrum where tissue optical absorption is lowest [4],



G. Liu, *Thermoplasmonics*, https://doi.org/10.1007/978-981-97-8332-8_5

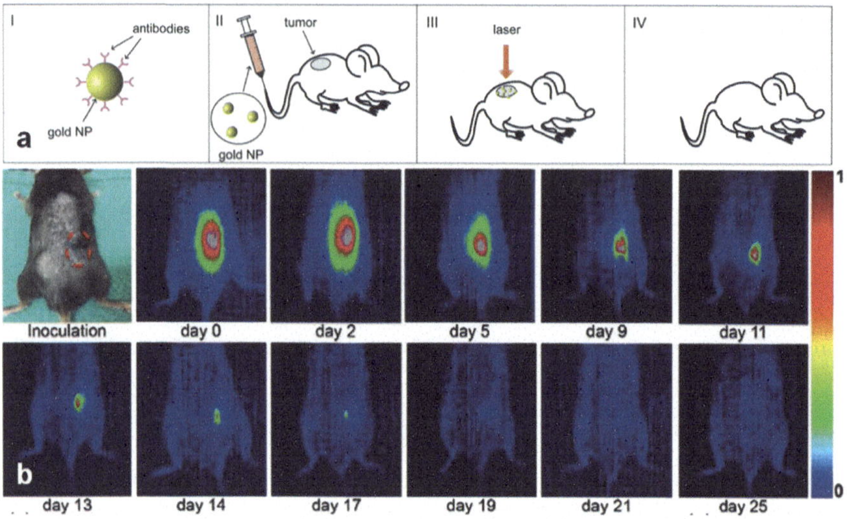


**Fig. 5.1** **a** Diagram of photothermal therapy. Initially, plasmonic particles are modified with either small molecules or antibodies specifically targeting cancer cells. These particles are then injected directly into the tumor site or intravenously via the tails of nude mice. Following an incubation period, the tumor area is exposed to light to heat the particles and induce hyperthermia. This process is repeated until healing is achieved. Reproduced with permission from Ref. [3]. Copyright 2012 Wiley–VCH. **b** Fluorescence imaging series showing a mouse injected with cancer cells and $Ag_2S$ nanoparticles, recorded over several days after inoculation. Reproduced with permission from Ref. [8]. Copyright 2020 Wiley–VCH

minimization of metal content to reduce toxicity, and careful selection of nanoparticle size and surface chemistry to maximize cellular uptake [5, 6]. Due to the effective light absorption, only relatively low input power is required to ablate tumors, enabling selective and efficient thermal destruction while sparing surrounding healthy tissue [7].

PTT for cancer treatment was initially proposed in 2003 [9]. This method involves the targeted delivery of plasmonic nanoparticles to cancerous tissues, followed by light exposure to raise the local temperature enough to kill cancer cells without harming adjacent healthy tissue. Numerous plasmonic particles have been explored for their efficacy as photothermal agents, covering a range of materials, sizes, and shapes [10–12]. Additionally, PTT has undergone clinical pilot trials, proving to be a safe and effective treatment. Clinically, PTT has achieved advancements in treating conditions like head and neck cancer, coronary atherosclerosis, diabetic and age-related macular degeneration, prostate cancer, and genitourinary syndrome in menopause [13–15]. Since the use of silica-cored Au nanoshells in clinical settings [16], an increasing number of inorganic agents are advancing into clinical trials [17]. For example, the $Ag_2S$ nanoheater equipped with multiparameter sensing offers reliable thermal feedback during live tumor therapy (Fig. 5.1b) [8, 18–20]. Following the clinical trials, PTT shows significant encouraging results, particularly in prostate cancer therapies involving both large animals and human patients [21, 22].

Despite the advancements, there are still several challenges to be overcome. One significant hurdle in PTT is efficiently delivering the particles to the tumor site [23, 24]. Ideally, delivery should aim to minimize the spread of particles to distant organs while ensuring even distribution within the tumor itself. There are essentially two main strategies for delivering particles: (i) Local injection, which encounters the challenge of requiring precise tumor localization. Achieving homogeneous distribution through this method is complicated owing to the increased internal pressure within the tumor, which can counteract the injection process. (ii) Utilizing the enhanced permeability and retention effect by taking advantage of the unique properties of tumor-associated blood vessels. These vessels, formed through the process of angiogenesis, differ significantly from normal tissue vessels [25] Notably, they function like a sieve, particularly effective for nanoparticles sized between 60 and 100 nm, allowing these particles to accumulate efficiently in the tumor site [26].

Another critical issue pertains to assessing its efficiency. Common methods for evaluation include measuring the survival times of treated animals relative to a control group, or using calipers or computer tomography scans to determine tumor size [23]. However, these size measurements may not reflect treatment efficiency since necrotic tissue might remain even as the tumor size decreases. Particularly with aggressive cancer, it is vital to target cancer stem cells responsible for recurrence. Merely measuring tumor volume fails to provide information on whether the cancer stem cells were effectively targeted. Positron emission tomography (PET) is gaining recognition as a powerful tool for evaluating the efficacy. PET enables noninvasive monitoring of cell viability within tumors, allowing clinicians to identify areas where viable cancer cells persist posttreatment [22, 27]. A notable advantage of PTT is that the nanoparticles remain in the tissue after treatment. This allows for the therapy to be rapidly and easily repeated using the same particles, enhancing treatment convenience and effectiveness.

A fundamental aspect of PTT requires deep understanding the mechanisms through which it induces cancer cell death, alongside the biological and physiological responses. Although a bulk of research using subcutaneous cancer models in mice has shown promising results, the applicability of these models to human biomedical conditions frequently remains uncertain [11, 28, 29]. To bridge this gap, there is an urgent requirement for the development of more sophisticated orthotopic models. These models involve implanting human tumors into the corresponding organs of animals, providing a more accurate simulation of tumor vascularization, as well as the potential impact on adjacent healthy tissues and vital organs. Besides, there is an increasing tendency to enhance the effectiveness of PTT by combining it with other treatments such as radiotherapy, chemotherapy, and immunotherapy [20, 30–32]. This approach aims for a synergistic effect that could potentially increase the overall success rates of therapies. Parallel to this, the same particles have also demonstrated efficacy and stability as photosensitizers. Under low-intensity illumination, these particles can produce reactive oxygen species, facilitating their use in photodynamic therapy.

### 5.1.2 Drug and Molecular Delivery

Drug and molecular delivery determine the therapeutic performance [15, 33]. In this context, plasmonic heating opens up innovative pathways for controlled release and targeted therapy. Plasmonic particles can be conjugated with specific drugs or molecules and directed toward a particular site in body [34, 35]. In delivery systems where drugs are either encapsulated or attached to the surface of nanoparticles, the use of plasmonic heating can initiate the release of these drugs. This release mechanism can occur either by melting a temperature-sensitive polymer layer or by loosening the chemical bonds that secure the drug to the nanoparticle [36]. Additionally, the process of nanoparticles being absorbed by cells enables the controlled release of the drug from within the cell's cytoplasm through irradiation with near-infrared light [23]. The wealth of knowledge on the attachment of molecules to metal, especially gold, surfaces has inspired a variety of strategies for the laser-induced thermal release of drugs and genetic materials from nanoparticles.

Thermoplasmonics in drug delivery utilizes two main approaches: (i) direct conjugation of drugs to nanoparticle surfaces, where irradiation-induced heating triggers release (Fig. 5.2a), and (ii) the use of thermosensitive carriers such as hydrogels or lipids that discharge contents upon heating (Fig. 5.2b) [37, 38]. Notably, specially designed nanoparticles remain inactive plasmonically until activated in $H_2O_2$ environments typical of tumor sites, thereby minimizing thermal effects in nontarget tissues [39]. Research also shows that short RNA strands bound by thiol bonds to gold surfaces are released via laser irradiation, believed to be through heat-induced dehybridization of the RNA [40]. Despite the prevalent theory of thermal melting initiating release, alternate studies highlight an electrochemical release mechanism via hot electrons [35]. Recent experiments have demonstrated that pulsed lasers specifically cleave gold-thiol bonds to release RNA duplexes, whereas continuous-wave laser exposure results in both single and double-stranded RNA releases [41]. This evidence underscores the dual role of thermal and electrochemical processes in the release of nucleic acids from gold nanoparticles.

A range of plasmonic structures, notably hollow nanostructures and 2D nanosheets known for high drug-loading capacity, have been developed as nanocarriers for controlled drug release [34, 43–45]. An interesting category within these includes gatekeeping particles, which feature a core built from porous inorganic materials that house the drug cargo, encapsulated by an outer shell that regulates cargo release [46]. For instance, a gold nanostar enveloped by a mesoporous silica shell topped with a paraffin layer was engineered for controlled drug delivery. Loaded with doxorubicin, the mesoporous silica enables release upon NIR laser irradiation through the melting of the paraffin coat, effectively diminishing the viability of human breast cancer cells [47]. Concurrently, studies on laser-induced drug release from plasmonic nanoparticles, such as those utilizing the hollow interiors of nanocages, further corroborate the potential of these platforms [44, 48, 49]. However, challenges remain in optimizing in vivo targeting and mitigating drug leakage from these nanoparticles in the bloodstream.

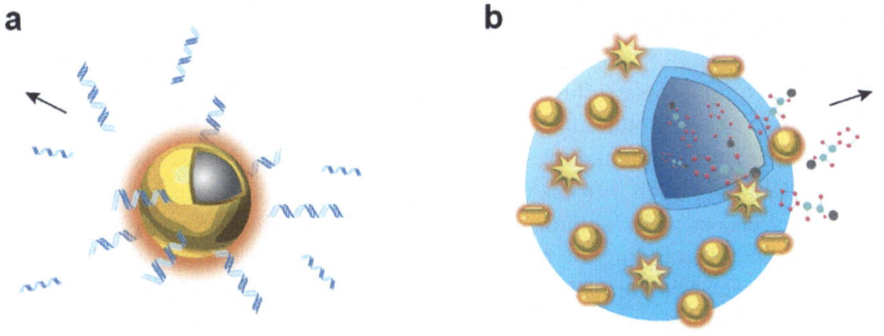


**Fig. 5.2** Plasmonic heating strategies in drug delivery: **a** thermally induced release of molecular payloads attached to nanoparticle surfaces, and **b** temperature-triggered changes in permeability within thermosensitive materials (e.g., liposomes and hydrogels) encapsulating drugs. Adapted with permission from Ref. [42]. Copyright 2024 American Chemical Society

Drug delivery can also be achieved by involving transient disruption of the cell membrane [50, 51]. This technique employs the rapid temperature increase induced by plasmonic pyramid arrays to disrupt cell membranes, facilitating intracellular transport. The process enhances membrane permeability by creating temporary pores that allow for the direct diffusion of miRNA molecules. To reduce toxicity toward healthy cells, an approach allowing intercellular delivery without requiring direct contact between cells and photothermal nanoparticles has been explored [52]. Utilizing biocompatible electrospun nanofibers embedded with iron oxide nanoparticles, visible-light absorption and efficient heat transfer are achieved directly to specific regions of the cell membrane, thereby increasing membrane permeability. Rapid temperature elevation at the interface between the photothermal nanostructure and the cell membrane leads to the formation of nanobubbles due to liquid environment vaporization [53, 54]. The dynamic expansion and collapse of these vapor nanobubbles generate high pressure and vigorous fluid flows, disrupting adjacent membranes and promoting intercellular delivery. Additionally, the nanobubbles act as a barrier, preventing thermal energy dispersion into the surrounding area and thus enhancing cellular viability. This membrane-disruption strategy presents a low toxicity, high-throughput route for delivering various biomolecules and nanoparticles into cells.

Ongoing research and development in nanomedicine are enhancing thermoplasmonic strategies to optimize drug and molecular delivery while minimizing associated risks [55]. Future directions in this field focus on advancing substrate engineering and developing smart materials, poised to broaden thermoplasmonic applications [56–58]. Materials with high thermal sensitivity are being explored as heat-triggered actuators both in cellular environments and in interfaces with biological specimens.

### 5.1.3 Cell Biology

All cellular components exhibit a high degree of temperature sensitivity: lipid layers can transition phase, proteins may unfold, and DNA strands can denature [2]. Precise temperature control is crucial for studying these phenomena in vitro. Selective laser heating of biological in vitro samples, using plasmonic nanoparticles as efficient light absorbers, provides significant advantages. Firstly, the reduced spatial scale made possible by coupling the heating laser with an optical microscope allows for micrometric area heating, enabling targeted heating of individual cells [59]. This approach facilitates the application of diverse temperature conditions within a single multiwall plate, even allowing for the heating of subcellular compartments or individual organelles to address fundamental questions in thermal biology. Secondly, the reduced timescale of microscale experiments ensures rapid temperature increases, approximately within 1 ms, establishing a precise initial time point for assessing the dynamic responses of biological systems, which may occur in under a second. Thirdly, the significant local temperature increases achievable with pulsed illumination can be utilized in applications like nanosurgery or optoporation, while continuous-wave illumination can also achieve temperatures that are unattainable with conventional macroscopic methods [60]. Thus, thermoplasmonic heating opens new opportunities for single-cell analysis and manipulation.

Plasmonic heating has facilitated both extra and intracellular manipulations, primarily enabled by efficient nanoparticle delivery via endocytosis. Beyond the traditional methods, thermoplasmonics offers an innovative alternative for nanoparticle delivery through optically triggered injection. Researchers demonstrated high cellular viability using effective delivery of gold nanoparticles, achieved by combining plasmonic heating with optical pushing at the membrane site to expedite particle entry through the lipid bilayer [61]. This thermoplasmonic perforation opens new avenues for in-situ single-cell surgery. Additionally, the manipulation and extraction of intracellular organelles from living cells were accomplished through transient light-induced membrane perforations, assisted by plasmonic heating in conjunction with optical trapping and micropipette techniques [62]. By adjusting optical power, researchers have managed both to induce membrane perforation and facilitate membrane repair, leveraging the variable lipid fluidities at high and mild temperatures, respectively (Fig. 5.3a). These advancements in thermoplasmonic membrane manipulation significantly enhance the capabilities of single-cell level interventions.

Thermoplasmonics can also modulate transcriptional regulation in single cells, indicating that particle heating guides the directed differentiation of dental pulp stem cells by regulating mitochondrial metabolism [63]. Furthermore, remote modulation of myotube [64] and neuronal cell differentiation [65] was achieved via plasmonic heating (Fig. 5.3b). Although the exact biological mechanisms remain elusive, evidence shows the thermal regulation of gene expression affects several proteins, including heat shock proteins and other stress-responsive proteins [64, 66]. Plasmonic heating has been widely applied to manipulate cellular behavior by thermally

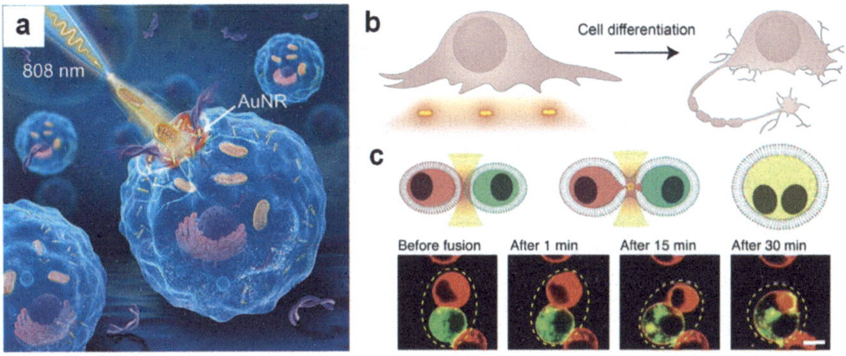


**Fig. 5.3** **a** Cellular perforation was achieved via irradiation of Au NRs close to the cell surface, facilitating subsequent intracellular manipulations through the integration of optical trapping with a micropipette. Reproduced from Ref. [62] Copyright 2021 American Chemical Society. **b** Neurite outgrowth was induced by plasmonic heating. Reproduced from Ref. [42]. Copyright 2024 American Chemical Society. **c** Photothermal fusion of living cells using an optically trapped gold nanoparticle, as demonstrated in fluorescence images. Reproduced from Ref. [72]. Copyright 2017 Tsinghua University Press

inducing changes in the extracellular environment. Embedding cells within thermally sensitive hydrogels enables researchers to toggle cellular connections on and off through localized heating, facilitating studies on cellular stimulus propagation within neuronal networks [67]. In experiments, neural cells were positioned atop a layer of gold nanorods, subject to localized laser-induced heating. The cells, separated by the thermosensitive hydrogel, allowed for controlled growth of selected connections through selective ablation of gel areas. Thermally induced hydrogel contraction has further been utilized to exert subcellular mechanical stress by embedding a contractile hydrogel between polymeric pillars, onto which cells adhere, leveraging this setup to apply mechanical stimuli [68].

Advanced experiments have been performed on living organisms, including bacteria and eukaryotic cells, to facilitate the precise fusion of select cells or vesicles [23, 69]. The irradiation of plasmonic nanoparticles localized at the juncture between two vesicles or cells induces immediate complete fusion. This fusion process is effective with both pulsed and continuous-wave lasers, provided the nanoparticles are proximal to the membranes of the targeted entities (Fig. 5.3c) [70–72]. Such methods offer substantial potential for generating hybrid cells, enabling controlled integration of proteins and buffers in biophysical settings, and conducting picoliter-scale chemical reactions through the fusion of vesicles loaded with specific reactants. A notable application involved the fusion of dendritic cells and Burkitt lymphoma B cells using pulsed laser activation of gold nanoparticles at the resonance frequency [71]. These Au NPs were bonded to antibodies that target specific cell surface receptors, facilitating selective fusion when exposed to laser light. This approach not only highlights the operational versatility of employing lasers with plasmonic nanoparticles but also underscores its utility in orchestrating complex biochemical processes in cellular contexts, such as initiating cell-free protein synthesis from a GFP plasmid through the merger of three separate vesicles [73].

Thermoplasmonic applications enable the study of cellular processes with enhanced spatial and temporal precision, marking the emergence of an expanding research domain potentially termed cell thermal biology. Despite their promising potential, the use of plasmonic heating in cellular biology faces significant challenges including nanoparticle cytotoxicity, the necessity for precise thermal management to avoid harming nontarget cells and tissues, and the need for advanced, costly equipment for targeted irradiation.

### *5.1.4 Polymerase Chain Reaction*

Polymerase chain reaction (PCR) serves as the benchmark molecular diagnostic technique for nucleic acid amplification and detection, attributed to its superior sensitivity and specificity. This technology operates through a triphasic thermal cycling process: (1) denaturation, wherein DNA strands separate at elevated temperatures; (2) annealing, during which complementary single-stranded DNA fragments hybridize at reduced temperatures; and (3) extension, where a new DNA strand synthesizes from the annealed template via polymerase action [74]. This mechanistic framework facilitates exponential replication of billions of target nucleic acid segments. Additionally, PCR allows for estimations of initial nucleic acid concentrations using cycle threshold values, reinforcing its status as the definitive standard for quantitative diagnostics across fields such as biology, medicine, and agriculture [75, 76]. The method necessitates recurrent cycles of heating and cooling to achieve dehybridization and rehybridization of target sequences. Standard thermocyclers, which utilize Peltier element-based heating blocks, generally accomplish the amplification within 1–2 h due to inherent limits in temperature adjustment rates [15]. Emerging methods like plasmonic heating are advancing the pace of DNA or RNA detection, offering significant improvements over traditional techniques.

Thermoplasmonic heating and cooling has given rise to photonic PCR, a method that accelerates PCR processes [77–80]. In this approach, the full potential of plasmonic heating is harnessed using both nanoparticles and substrates to drive the thermal cycling necessary for DNA amplification. Photonic PCR overcomes the time-intensive thermal cycling constraints of conventional PCR through faster temperature transitions enabled by plasmonic materials. Initial demonstrations involved the use of thin gold films for optical heating [81], with subsequent adaptations such as an Au-nanorod-based PCR system designed for rapid bacterial detection, integrating cell lysis and DNA amplification into a single accelerated step (Fig. 5.4) [82]. This innovation significantly reduces analysis time to under 20 min. Further enhancements are achievable by refining the structural properties of nanomaterials to boost photothermal conversion [83], thus minimizing thermocycling durations, as evidenced in 2017 when ultrafast cycling, accomplishing 30 cycles in 54 s, was demonstrated using Au nanorods heated by a NIR resonant laser [84]. These advances

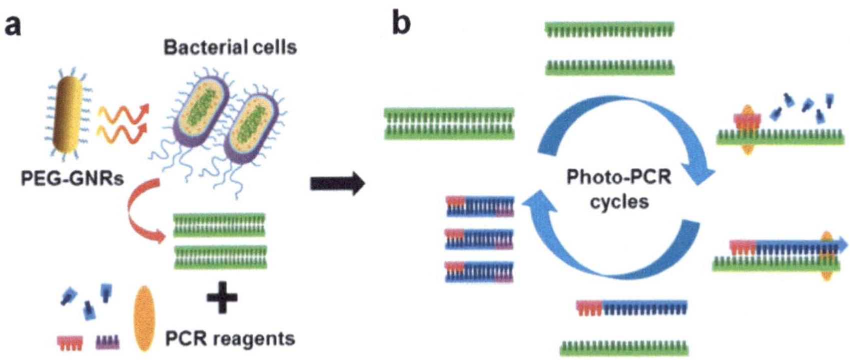


**Fig. 5.4** Plasmonic heating via gold nanorods facilitates bacterial lysis and DNA extraction for photonic PCR in a singular process. **a** A diagram illustrating the components involved in the photonic PCR reaction and **b** the amplification sequence of bacterial DNA. Reproduced with permission from Ref. [82]. Copyright 2017 Ivyspring International Publisher

pave the way for the development of efficient point-of-care applications, potentially reducing both cost and operational times. Continual technical enhancements could lead to even quicker thermocycling in photonic PCR, though enzyme kinetics constrains the reaction speed.

### *5.1.5 Fight Against COVID-19*

The COVID-19 caused by the severe acute respiratory syndrome coronavirus 2 (SARS-CoV-2) has precipitated a global pandemic that poses significant threats to public health. Researchers globally have achieved substantial progress in developing personal protective equipment, effective vaccines, reliable diagnostic tools, and therapeutic strategies [85, 86]. Nanomaterials, characterized by their diminutive size, highly reactive and adjustable surfaces, low toxicity, and potential for chemical customization, represent a burgeoning avenue for enhancing point-of-care diagnostics, therapeutic delivery systems, and antibody development with tailored nanodevices aimed at detecting, treating, and curtailing the spread of SARS-CoV-2 [87]. Furthermore, plasmonic nanoparticles have introduced new opportunities for cost-effective and scalable detection and treatment modalities [88]. These innovations include the use of plasmonic nanoparticles in (i) preventive measures and disinfectants, (ii) diagnostic assays for rapid, straightforward, sensitive, and specific identification, and (iii) delivery systems for antiviral agents in therapeutic and vaccine applications.

Beyond PCR testing, respiratory face masks may be enhanced with photothermal nanomaterials for antimicrobial efficacy. Superhydrophobic and thermoplasmonic coatings on these masks have shown superior protection compared to existing devices

[89–91]. Plasmonic heating can increase the surface temperature of the mask to 80 °C within a minute under solar exposure, while its hydrophobic properties prevent respiratory droplets from lingering on the surface. The synergistic integration of photothermal and photocatalytic effects grants these masks reusability and self-sterilizing capabilities (Fig. 5.5a) [91]. Moreover, silver nanoparticles embedded in the masks provide additional microbial protection through ion disinfection [92]. These combined effects not only bolster personal protection but also encourage the development of more effective protective gear to counteract COVID-19. Further advancements have been made in creating various reusable and antimicrobial face masks using diverse photothermal nanomaterials.

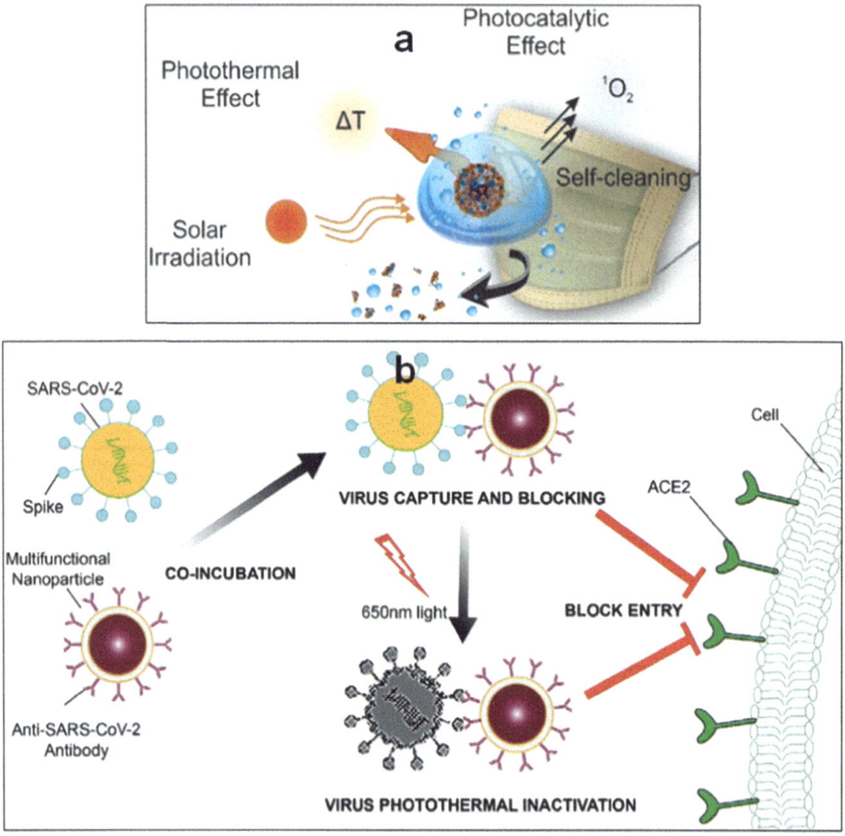


**Fig. 5.5** **a** Virus inactivation in a nonwoven surgical mask achieved through photothermal, photocatalytic, and hydrophobic self-cleaning mechanisms. Reprinted with permission from Ref. [91]. Copyright 2021 American Chemical Society. **b** Schematic of a multifunctional photothermal particle designed for the capture (via antibody) and inactivation (through photothermal conversion) of SARS-CoV-2. Reprinted with permission from Ref. [93]. Copyright 2021 Advanced Science published by Wiley–VCH

Moreover, functionalizing plasmonic particles with high-affinity neutralizing antibodies targeting SARS-CoV-2 spike proteins has been engineered to efficiently capture the virus (Fig. 5.5b) [93]. Such antibody-mediated interception prevents SARS-CoV-2 from entering host cells, while subsequent inactivation is achieved through photothermal effects. This method of imbuing photothermal nanoparticles with antiviral properties presents significant promise for intercepting SARS-CoV-2 and mitigating infection. Plasmonic particles can enhance pharmacokinetic profiles, offering insights that could enhance antiviral treatments against COVID-19 and potential future variants [86]. While these strategies are in preliminary phases, they need further refinement for expanded use in preclinical and clinical evaluations to ascertain the efficacy of the nanoparticles against COVID-19.

## 5.2 Optoelectronic Information Science

### *5.2.1 Thermoplasmonic Sensor*

The detection of gases and chemical reactions is essential for both scientific and industrial applications [94, 95]. To facilitate the use of smaller reagent concentrations and sensing volumes, optical sensors have shifted from thin-film systems to engineered nanodevices [96, 97]. In such nanoscaled environments, plasmonic particles and nanostructures play a critical role in enhancing biosensors owing to their capacity to focus light into sub-wavelength electromagnetic hot spots and convey subtle environmental changes as modulations in their optical response [98, 99]. Predominantly composed of metallic nanoparticles or nanostructures, plasmonic materials are readily combined with various active materials to explore processes ranging from hydrogen absorption in palladium to the detection of trinitrotoluene [100–102].

Plasmonic biosensors exhibit four principal structure–property relationships pivotal for their functionality [98, 103]. (1) Surface-enhanced Raman scattering (SERS) leverages the potent electric fields surrounding plasmonic nanostructures to amplify Raman scattering signals from adsorbed molecules, offering a method characterized by high sensitivity and selectivity for molecular detection via unique vibrational signatures. The method's sensitivity hinges mostly on both the enhancement factor and the analytical enhancement factor [99, 104]. (2) Plasmonic-enhanced fluorescence facilitates significant fluorescence signal amplification through bridging the molecular absorption and emission spectra with the localized surface plasmon resonances of metal nanoparticles and fluorophores [105, 106]. (3) Colorimetric sensing detects analytes or environmental changes via the colorimetric response of colloidal nanoparticles. This method relies on nanoparticle aggregation or shape alteration in response to specific stimuli, resulting in distinct alterations in their optical characteristics [107]. (4) Chemiresistor sensing mechanism operates on resistivity changes in chemically sensitive Au nanoparticle films upon analyte adsorption, demonstrating high sensitivity for detecting both liquid and gas analytes [108]. The selection of thiols on the nanoparticle surface is crucial for enhancing the sensitivity and selectivity of Au nanoparticle-based chemiresistors [109, 110].

The unique of thermoplasmonics to generate heat within confined volumes facilitates the development of highly sensitive sensors [106, 111, 112] Localized surface plasmon resonances in nanoparticles induce both localized heating and significant enhancement of the local electromagnetic field. Such localized heating can catalyze chemical reactions or change the physical properties of a sample, enabling the detection of substances at very low concentrations [113]. Moreover, the signal from the target analyte is amplified due to the enhanced electromagnetic field, increasing sensor sensitivity. Precise control and localization of plasmon-induced heating permit spatially targeted activation of chemical or biological reactions. For instance, a dual-functional plasmonic biosensor utilizing both the photothermal effect and LSPR sensing has been developed for sensitive and reliable COVID-19 detection (Fig. 5.6) [114].

This sensor employs localized photothermal heat to increase the hybridization temperature of nucleic acids, enhancing the differentiation accuracy between similar gene segments, with a detection limit for specific SARS-CoV-2 sequences as low as 0.22 pM. Furthermore, by employing plasmonic nanoparticles of varying materials or shapes, which respond to different light wavelengths, multiplexed detection of various analytes can be simultaneously achieved through selective plasmon heating [115, 116]. Owing to their versatility and superior sensing performance, plasmonic

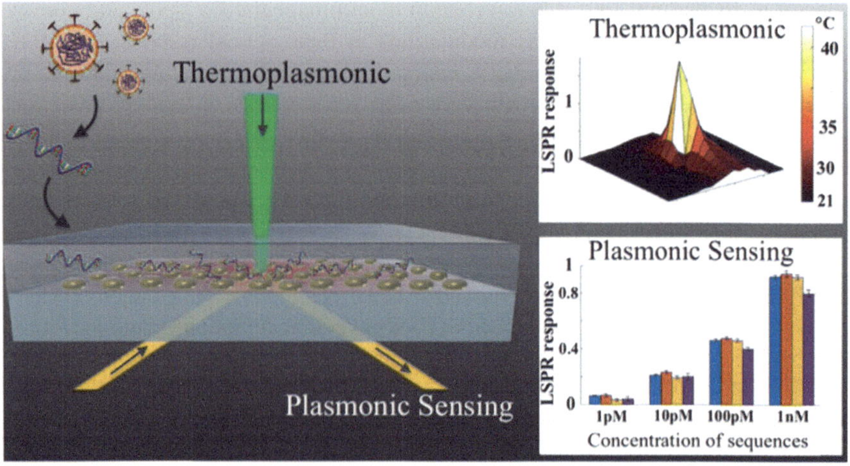


**Fig. 5.6** Thermoplasmonic biosensor. Left: Schematic depicting photothermally enhanced LSPR sensing. Right: Temperature distribution map surrounding the heat source on an Au nanoisland (top) and concentrations of various viral oligonucleotides measured with the dual-functional LSPR biosensor (bottom). Reprinted with permission from Ref. [114]. Copyright 2020 American Chemical Society

structures hold promise for detecting chemical and catalytic processes at the single-molecule level [117, 118]. As research progresses, the incorporation of innovative materials, technologies, and analytical approaches is set to broaden the scope of plasmonic sensing [98, 119, 120]. These advancements will facilitate diverse applications in biomedical sensing, environmental monitoring, and chemical analysis.

### 5.2.2 *Photothermal Imaging*

The photothermal effect leverages nonradiative plasmon decay to generate thermal contrast for imaging nanoparticles, termed photothermal imaging (PTI). This technique mainly employs two strategies: (i) photothermal interference contrast imaging (PIC) [121, 122] and (ii) photothermal heterodyne imaging (PHI) [123], both of which are pump–probe methodologies requiring dual beams—a heating beam and a probe beam (Fig. 5.7) [124, 125]. When the localized surface plasmon resonance of the nanoparticle is activated by the heating beam, it increases the temperature of the nanoparticles and their surrounding environment, altering their refractive index. This change in refractive index, induced by the photothermal effect, is detected using a nonresonant laser source known as a probe beam [126]. In this arrangement, both beams are co-aligned and concentrated on the same sample area through a microscope objective lens. Changes detected by the probe beam are captured via a photodiode connected to a lock-in amplifier, enhancing the signal-to-noise ratio [124]. PTI can further optimize signal sensitivity and noise reduction by selecting a probe beam wavelength at which the nanoparticles exhibit minimal absorption [127, 128].

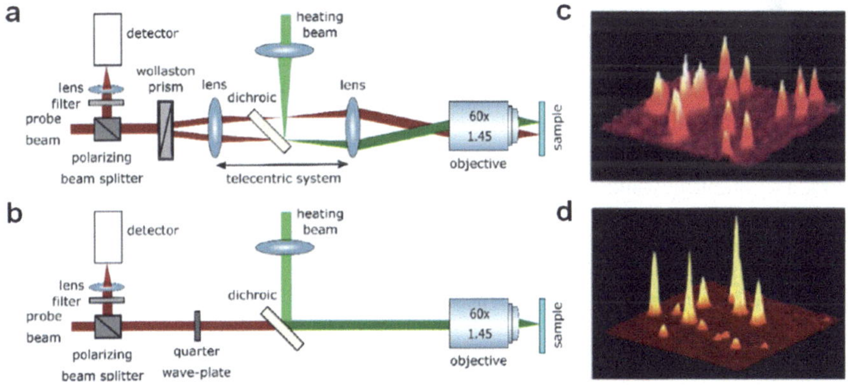


**Fig. 5.7** Experimental setup for: **a** Photothermal interference contrast and **b** Photothermal heterodyne imaging. **c** Photothermal image of single 5 nm gold nanospheres captured utilizing the PIC method. **d** Photothermal image highlighting single 2 nm (short peaks) and 5 nm (tall peaks) gold nanospheres obtained via the PHI method. Adapted from Ref. [129]. Copyright 2008 American Chemical Society

Boyer et al. pioneered the far-field detection of nanoscale metal particles using a photothermal technique that integrates high-frequency modulation with polarization interference contrast (Fig. 5.7a, c) [121, 122]. This method captures the absorption signal by quantifying the subtle phase shift in a probe beam resulting from the thermal effect on a metal nanoparticle. By employing a 514 nm heating laser, modulated at a high frequency ranging from 100 kHz to 1 MHz via an acousto-optic modulator, this approach significantly enhances signal-to-noise ratio by filtering out low-frequency noise. Alternatively, Berciaud et al. introduced a more straightforward and sensitive photothermal heterodyne imaging (PHI) technique that relies on a single probe beam path, in contrast to the dual paths utilized in photothermal interference contrast imaging (Fig. 5.7b, d) [123, 129]. This streamlined configuration circumvents the challenges of depolarization and residual imbalances inherent to the dual-arm PIC interferometer setup. Here, the same or reflected probe beam serves as a reference, producing interference between the scattered probe and reference beams akin to the interferometric scattering method [130]. The reference beam may either be the reflection from the sample-coverslip interface or the transmitted beam through the sample.

A probe-free detection approach utilizing a single laser beam was developed [131]. This single-laser configuration obviates the need for an additional laser and does not rely on meticulous alignment of pump and probe beams. Here, a singular modulated laser beam fulfills both pump and probe functions. Interference arises between the thermally scattered probe field and the transmitted probe field, with the resultant signal captured using a lock-in amplifier. The out-of-phase component of this signal indicates the presence of an absorbing object. Although this method achieves substantial contrast for absorbing objects within environments of strong scatterers, its signal-to-noise ratio is significantly lower than that observed in photothermal heterodyne imaging by at least an order of magnitude. To date, a variety of plasmonic nanoparticles have been utilized in photothermal imaging for biomolecular visualization [132–134]. Yet, practical applications demand precise design and synthesis of nanostructures with optimized plasmonic functionality specific to the intended use. For enhanced bioimaging efficacy, PTI should be integrated into practical multimodal imaging strategies, such as combining with optical fluorescence, MRI, CT, PET, and SPECT—enabling visualization of anatomical structures and tumor locations [135].

### 5.2.3 *Photoacoustic Imaging*

Photoacoustic (PA) imaging is a novel modality in which pulsed laser light generates detectable pressure transients via conventional ultrasound technology [136]. This technique merges the benefits of optical and ultrasonic methods, facilitating high-resolution detection of deep tumors. Plasmonic nanoparticles, including gold nanorods and nanostars, are commonly employed as PA contrast agents [137–140]. These particles rapidly heat upon light absorption, leading to thermoelastic expansion that produces pressure waves in the surrounding medium. These acoustic waves are

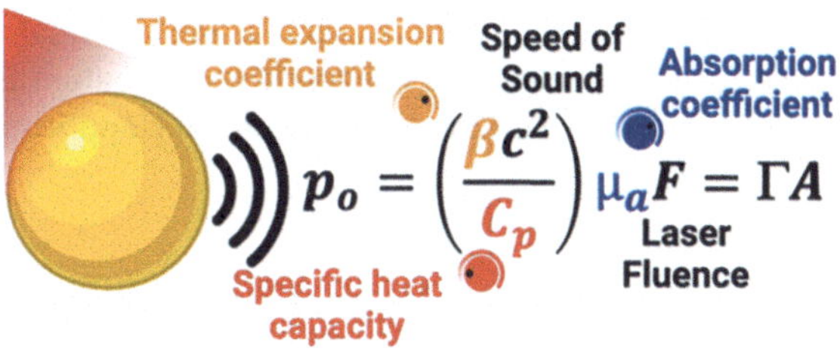


**Fig. 5.8** Thermoelastic expansion model. Adapted from Ref. [141]. Copyright 2020 American Chemical Society

then detectable by external ultrasound detectors. The process is effectively described by the thermoelastic expansion model (Fig. 5.8) [141]. Light absorption slightly increases temperature, causing thermoelastic expansion and consequent conversion of optical energy into pressure waves ($\boldsymbol{p}_0$), influenced by several factors: (i) absorption coefficient ($_a$), (ii) thermal expansion coefficient ($\boldsymbol{\beta}$), (iii) specific heat capacity ($\boldsymbol{C}_\mathbf{p}$), (iv) sound speed in the medium ($\boldsymbol{c}$), and (v) illumination fluence ($\boldsymbol{F}$). Adjustments to geometry, composition, coatings, and solvent environments around the plasmonic nanostructures can optimize these variables to enhance PA signal output.

Plasmonic particles can be tailored to exhibit peak absorption within the near-infrared region, effectively bypassing the higher absorption rates of biological tissues. This characteristic allows for enhanced light penetration and more robust photoacoustic signals from deeper tissue layers. For instance, a novel NIR-II plasmonic theranostics system leveraging silica-encapsulated gold nanochains has been developed, facilitating precise tumor diagnostics [142, 143]. The unique, transformable structure of these nanochains extends the absorption spectrum into the NIR-II region through the fusion of adjacent gold nanoparticles within the confined space of the chains, thereby improving photoacoustic imaging capabilities for tumor assessment. Moreover, surface-functionalized nanoparticles offer additional molecular targeting options through specific binding moieties [139, 144, 145]. Although thermal expansion remains the primary mechanism for generating photoacoustic transients with these agents, recent innovations have introduced exogenous contrast agents that utilize vaporization to enhance photoacoustic signal generation, significantly surpassing signals derived from thermal expansion alone [146, 147]. These agents, known as photoacoustic nanodroplets, are composed of liquid perfluorocarbon nanodroplets encapsulating plasmonic nanoparticles. Upon irradiation, a heat-induced liquid-to-gas phase transition generates substantial photoacoustic transients for improved contrast [146]. Furthermore, different materials and shapes of nanoparticles exhibit different LSPR peaks, enabling the development of multiplexed imaging strategies where multiple targets can be imaged simultaneously with high specificity using multispectral photoacoustic imaging [148–150]. This allows researchers to study complex biological interactions and disease processes in vivo.

Thermoplasmonics play a critical role in photoacoustic imaging by enhancing the contrast and resolution of imaging. This not only enables the visualization of deep tissue structures but also opens doors to functional and molecular imaging with potential impacts on disease diagnosis and treatment [39, 141, 151]. However, challenges in this field remain, such as the potential toxicity of nanoparticles, the need for precise control over nanoparticle distribution and the optimization of particle size, shape, and material for specific imaging applications.

### 5.2.4 Heat-Assisted Magnetic Recording

Enhancing data storage density is crucial for advancing cost-effective digital technologies. The predominant form of storage within cloud computing involves magnetic hard disk drives, which face an increasing demand for greater recording density, currently expanding at a rate of 15–20% annually [152, 153]. Addressing the trilemma of signal-to-noise ratio, thermal stability, and recordability is essential for further increases in recording density [154, 155]. The advent of heating-assisted magnetic recording (HAMR) emerges as a significant innovation in data storage technology. HAMR employs near-field transducers to locally and temporally elevate the temperature of a sub-diffraction-limited area in the recording medium above its Curie temperature. This reduction in magnetic coercivity enables the usage of smaller grain volumes while maintaining data's thermal stability, thereby enhancing storage density in forthcoming generations of hard disk drives [156–158].

The essential component of HAMR is the near-field transducer (NFT), which applies heat via a laser onto the recording medium. Variations exist in the design of the light delivery system to the NFT based on specific implementations [156, 159]. Generally, the HAMR head includes a laser diode operating near 800 nm wavelength, coupled to a waveguide through a mode coupler (Fig. 5.9a). The light advances down the waveguide and couples into the NFT, which subsequently emits the light into the recording medium, converting it to thermal energy. Waveguide configurations may include a parabolically shaped guide, a planar solid immersion mirror with a dual offset grating to focus the waveguide mode [156], a thin-film dielectric waveguide with a high refractive index core [159], or a metallic surface plasmon waveguide [160]. The optical spot emitted by the NFT must be sufficiently localized to achieve areal densities ranging from 1–5 Tb/in$^2$. At a bit aspect ratio of 4.0 for 1 Tb/in$^2$, the track density stands at 500 ktpi—approximately 16 times finer than the wavelength of the diode laser (around 800 nm) [159].

NFT design primarily revolves around the excitation of surface plasmons in a nanostructure, which reradiate to produce a sub-diffraction-limited light spot. Advanced integrated designs leverage waveguides to direct light onto custom-fabricated metal nanostructures, enabling isolated tracks to be written with bit spacings as narrow as 50 nm on an unpatterned ferromagnetic layer, or even 28 nm on a lithographically patterned metal islands layer (Fig. 5.9b) [156, 159]. The amplified field induced by the plasmonic nanostructure is confined to the near field with substantial divergence, yet this poses no issue in hard disk drives, given the mere nanometer-scale separation between the head and the medium during operation [154]. Plasmon heating-assisted magnetic recording promises to expand the frontiers of data storage technology, facilitating the development of smaller, more efficient, and

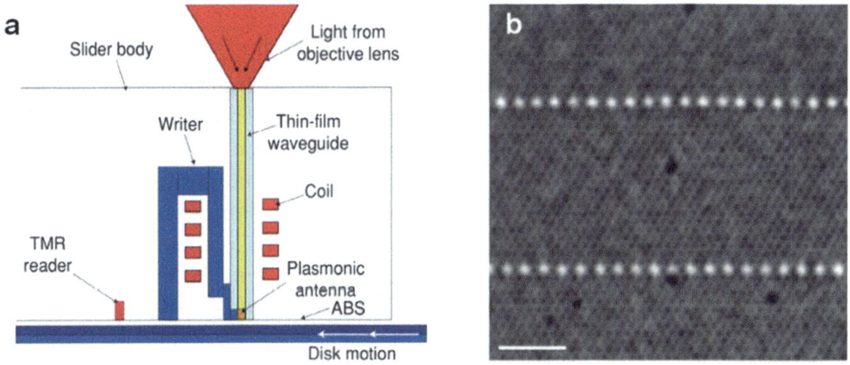


**Fig. 5.9** **a** Writing head for thermally assisted magnetic recording employing plasmon-enhanced near fields. **b** Magnetic-force-microscopy image displaying tracks in a cobalt/palladium disk array written using the head, featuring a 200 nm scale bar and a 28 nm bit length. Adapted with permission from Ref. [159]. Copyright 2010 Nature Publishers Ltd

higher-capacity storage devices. This advancement has far-reaching implications for data centers, cloud storage, and consumer electronics.

### 5.2.5 *Thermoplasmonic Infrared Detector*

Photodetectors find applications across diverse fields such as spectroscopy, imaging, flame detection, water purification, and biotechnology [161, 162]. Contemporary photodiodes often experience optical losses and suffer from charge-carrier recombination [163]. Plasmonic nanostructures are garnering attention in photodetection owing to their potential for ultrafast response and enhanced sensitivity. These structures localize electromagnetic fields, concentrating light near metallic surfaces or nanostructures, thereby enhancing the absorption and generation of hot carriers—electrons and holes far from lattice temperature equilibrium. The hot carriers produced by the nonradiative decay of surface plasmons in these nanostructures can be injected into a semiconductor's conduction band, enabling sub-bandgap photodetection [164–166]. Moreover, by manipulating the interfacial barrier height between plasmonic materials and semiconductors/insulators, hot-carrier photodetectors have shown significant advancements for broad-spectral-range (ultraviolet to infrared) applications.

Hot-carrier photodetectors, formed by a Schottky barrier with a semiconductor, exhibit unique properties such as sub-bandgap photodetection, straightforward fabrication, self-powering, room-temperature operation, and high tunability in terms of bandwidth, working wavelength, and polarization-dependence [167, 168]. The barrier height within these detectors can vary significantly, from approximately 3.8 eV (e.g., tin/silicon dioxide, $Sn/SiO_2$) to 0.3 eV (e.g., platinum silicide/p-type silicon, PtSi/p-Si), influenced by the material composition of the junction as well as fabrication methods and surface treatments [169, 170]. These photodetectors are capable of operating across a broad spectral range, from ultraviolet to terahertz, demonstrating their versatility in the electromagnetic spectrum [171–174]. For instance, hot electrons in the Sn-$SiO_2$-Si system with a 3.8 eV barrier height facilitate solar-blind UV detection [175], while plasmonic structures like gold-aluminum oxide-gold (barrier height ~ 2.6 eV) and gold-titanium dioxide-indium tin oxide (barrier height ~ 0.68 eV) enhance photoresponse in the visible-to-near-infrared ranges via surface plasmon resonance excitation [176]. Finally, utilizing silicon-based hot electrons can serve telecommunication photodetection, providing a cost-effective alternative to indium gallium arsenide (InGaAs) and germanium (Ge) detectors [168], while resonant THz photodetection is achieved using antenna-coupled graphene transistors exploiting graphene plasmons [177].

Despite their benefits, hot-carrier photodetectors face limitations in practical application due to low photoconversion efficiency [167, 178]. Additionally, plasmonic heating plays a dual role in photodetector operation, with both beneficial and detrimental impacts. Positively, plasmonic heating boosts hot-carrier generation, enhancing device efficiency and affecting the relaxation dynamics of these carriers, which can extend carrier lifetimes and hasten response times. Conversely, the increased temperatures cause thermal energy losses and induce unwanted thermal noise, degrading the signal-to-noise ratio. However, employing plasmonic-induced hot carriers is advantageous for infrared detectors where heat generation is favourable [179, 180]. One effective strategy involves coupling infrared radiation with nanomechanical resonators through nanoantenna-enhanced thermoplasmonic effects (Fig. 5.10). Implementing a hybrid nanoantenna-coupled mechanical device as a thermal IR detector has achieved a current responsivity of 12 mA/W, a displacement responsivity of 98.7 μm/W, and a thermal time constant of 5.7 ms at a wavelength of 6 μm [180]. This method demonstrates potential for expansion across various mechanical resonators, paving the way for new optomechanical sensing techniques.

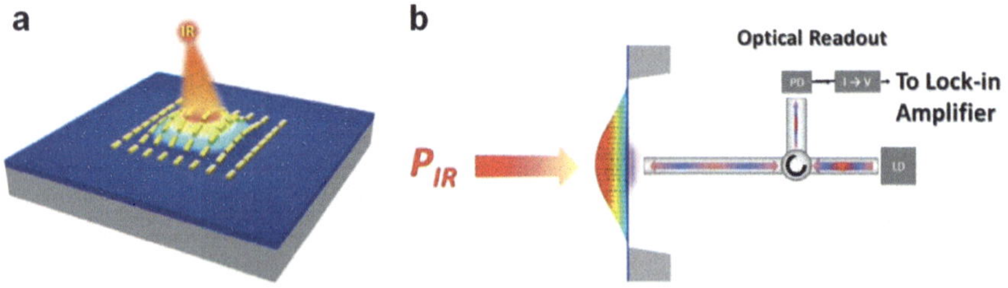


**Fig. 5.10** Plasmomechanical infrared detector: **a** The plasmonic antenna absorber consists of gold nanorod antennas fabricated on a commercial silicon nitride membrane. **b** this membrane serves as one of the two reflectors in a fiber optic interferometer. Adapted with permission from Ref. [180]. Copyright 2014 American Chemical Society

Advancements in materials engineering, including the exploration of low-loss plasmonic materials and engineered semiconductor heterostructures, are expected to address current challenges in hot-carrier photodetection [164]. The promise of 2D materials such as graphene and transition metal dichalcogenides, which possess unique electronic and optical properties, is increasingly recognized in this domain. Further innovations in nanostructuring techniques are poised to enhance both the performance and integration of these devices.

## 5.3 Energy Science

### *5.3.1 Thermophotovoltaics*

Photovoltaics facilitate the direct conversion of solar energy into electricity, playing a vital role in the energy sector. The performance of photovoltaic systems hinges on their capacity for light absorption, charge-carrier accumulation, and exciton dissociation [181]. Traditional photovoltaics are bounded by the intrinsic Shockley–Queisser efficiency limit due to thermal relaxation and losses from sub-bandgap photons [182, 183]. Plasmonic materials enhance cell performance through effective light trapping, energetic carrier generation, and resonance energy transfer [184–188]. This enhancement typically results from the synergistic interaction between plasmonic components and semiconductor layers. Light scattering increases the pathway of light, boosting absorption. Energetic carriers overcome the Schottky barrier at the metal–semiconductor interface, generating additional free electrons in the semiconductor's conduction band, thereby achieving high absorbance across broad wavelengths [189]. A third enhancement involves resonance energy transfer between the semiconductor and plasmonic component via dipole or multipolar coupling. Further details on these mechanisms are available in the referenced literature [185, 187, 190].

Thermophotovoltaics (TPVs) represent a strategy to minimize thermal losses by converting all incident electromagnetic energy into heat, which is then utilized locally [191–193]. In TPVs, an intermediate element absorbs sunlight, heats up, and generates thermal emissions specifically tailored to the bandgap of the solar cell, thereby enhancing its efficiency. Commonly, a TPV system comprises three essential components: a solar absorber that transforms solar energy into heat, a thermal emitter that reradiates this heat as infrared (IR) radiation, and a photovoltaic cell that converts this IR emission into electron–hole pairs to generate electrical current (Fig. 5.11). The overall efficiency denoted as $\eta_{stpv}$ is calculated as the product of the optical efficiency for concentrating sunlight $\eta_0$, the thermal efficiency of converting sunlight to heat at the emitter $\eta_t$, and the efficiency of electrical power generation from thermal emissions $\eta_{\text{tpv}}$ [194, 195]:

$$\eta_{\text{stpv}} = \eta_0 \eta_t \eta_{\text{tpv}} \tag{5.1}$$

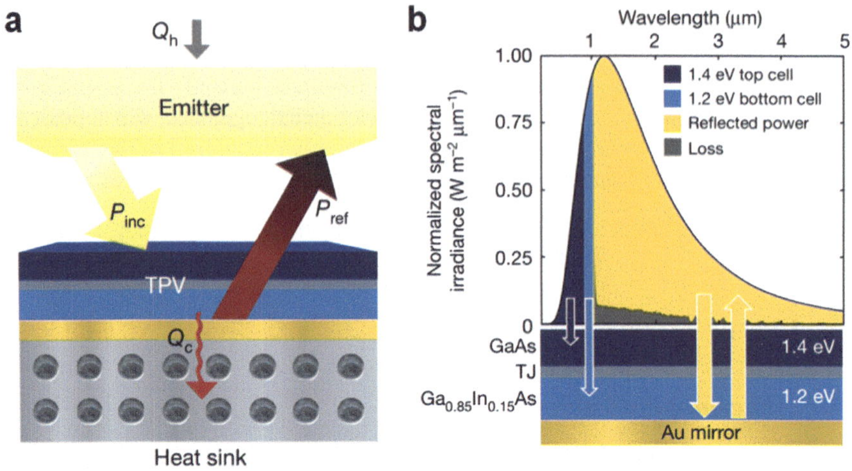


**Fig. 5.11** Tandem TPVs involve a dual-layer configuration **a** where solar energy incident on the TPVs is initially converted to electricity or reflected back to the emitter via a back reflector. **b** This system features spectral utilization in a GaAs-based tandem cell, which has demonstrated a record efficiency of 41%. Adapted with permission from Ref. [196]. Copyright 2022 Nature Publishers Ltd

Achieving a high-performance TPV hinges on the design of a solar absorber with efficient broadband solar absorption and a thermal emitter with narrowband emission that align spectrally with the PV cell's bandgap.

Achieving high efficiencies in TPVs requires addressing several challenges. Firstly, the design of highly efficient absorbers is critical to effectively convert absorbed solar energy into heat. Secondly, these visible-light absorbers must be paired with infrared thermal emitters capable of generating thermal radiation suitable for a photovoltaic cell with a low bandgap around 0.8–1.1 eV, which corresponds to an emitter temperature range of 1000–2000 K. This temperature requirement significantly constrains material selection, promoting refractory metals as prime candidates. For instance, photonic crystals made from titanium nitride, stable up to 1400 °C, are proposed for constructing emitter structures [192, 197, 198]. Spectral selectivity, either high or low absorption, in these structures must be strategically designed specific to the system to optimize cell efficiency. In practical implementations, tungsten thermal emitters and metasurface absorbers have been developed; these intermediate structures demonstrate high absorption in the visible-to-near-infrared spectrum while significantly reducing radiative losses at longer wavelengths, achieving efficiencies as high as 18.5% [192]. More recently, TPV systems have surpassed 40% efficiency using expensive multilayer structures with multiple bandgaps, complemented by highly reflective back surface reflectors (Fig. 5.11b) [193, 196].

Another promising approach involves the exploitation of near-field TPVs, which exhibit a substantial enhancement of thermal radiation transfer between the emitter and the photovoltaic cell as their separation is significantly reduced [2, 182, 199]. This effect arises due to optical tunneling across the narrow gap when it approaches the order of the radiative wavelength, creating an evanescent thermal near-field that increases energy transfer rates. This phenomenon of resonant energy tunneling exceeds the limitations dictated by the Stefan–Boltzmann law by several orders of magnitude, thereby facilitating a higher conversion of heat to electricity by solar cells. Despite a growing number of theoretical analyses on near-field TPVs employing plasmonic structures, experimental validations remain relatively scarce [200–202]. Nevertheless, the exploration of plasmonic materials within TPV design is driving research efforts aimed at leveraging plasmonic-induced heating to enhance both the efficiency and the adaptability of TPV systems [187, 203, 204]. Innovations in materials science, nanostructuring techniques, and system configurations are crucial in overcoming current obstacles, to realize the full potential of plasmonic-enhanced TPV technologies in practical applications.

### *5.3.2 Thermoplasmonic Desalination*

Containing over 90% of Earth's water [205, 206], seawater presents a solution to global freshwater scarcity via desalination techniques like multistage flash distillation and reverse osmosis [207, 208]. Despite significant energy inputs and financial costs associated with these methods, the abundant availability of solar radiation has catalyzed interest in solar-powered desalination strategies, particularly solar interfacial evaporation [209–211]. Solar interfacial desalination operates on three fundamental processes: (i) efficient conversion of solar radiation to thermal energy by a solar absorber, (ii) steam generation using this thermal energy, and (iii) condensation of steam into freshwater. Evaporation efficiency, which measures the ratio of thermal energy in vapor form to incoming solar flux, is critical in evaluating the performance of these systems

$$\eta = \frac{\dot{m} h_{fg}}{q_{\text{solar}} A} \tag{5.2}$$

here $\dot{m}$ represents the evaporation rate, $h_{fg}$ denotes the latent heat of vaporization, and $A$ indicates the absorber area. Enhancements in both evaporation rate and efficiency constitute a primary objective in desalination research, with current-state-of-the-art efficiency nearing 94%.

Solar desalination needs efficient conversion of solar photons into thermal energy by absorbers for water evaporation. Plasmonic absorbers have garnered significant interest due to their capacity to focus solar energy into compact volumes. The localized surface plasmon resonance effect generates a confined electric field that is converted into localized heat, thereby facilitating the evaporation of water near the plasmonic structures [212]. Initial studies of vapor generation focused on bubble formation from single nanostructures [213–215]. Later research explored the synergistic effects of multiple particles, which create numerous hot spots, aiming to achieve broadband solar absorption [216]. However, the photothermal efficiency and stability of these individually dispersed structures in solutions are compromised due to the reduced effectiveness of the collective synergy over extended distances between particles.

An alternative method to enhance absorption employs particle coupling, which requires the proximity of plasmonic structures in a complex arrangement of sizes and shapes, such as clusters or assemblies within porous frameworks. Drawing inspiration from natural systems, surfaces incorporating floating plasmonic films have been developed at the air/water interface, focusing thermoplasmonic heating at the water surface to facilitate phase change evaporation and minimize in-situ energy losses [217, 218]. Absorbers comprising film-supported gold, silver, or aluminum nanomaterials have achieved efficiencies exceeding 90% in solar energy capture for desalination [219–225]. Specifically, aluminum particles dispersed throughout the pores induce a hybridized plasmon resonance covering the entire solar spectrum (Fig. 5.12), resulting in an evaporation efficiency above 90% [219]. Furthermore, plasmon-decorated natural woods have demonstrated high conversion efficiencies in steam generation [226]. These film absorbers are cost-effective compared to other alternatives such as nanofluids and exhibit synergistic effects in collective heating, with the capability to float spontaneously on the water surface [227].

Solar desalination necessitates advancements at both material and system levels [228]. Recent inquiries have integrated steam generation with fuel formation under solar irradiation, achieving the conversion of cellulosic raw materials into bioethanol. This validates the potential for off-grid solar methods in sustainable bioethanol production [229]. Despite the high cost of plasmonic materials, there is a compelling need to develop economical and efficient solar–thermal materials for widespread application [210]. Historically, most research has been conducted under laboratory conditions using brine as a surrogate for seawater, with minimal focus on

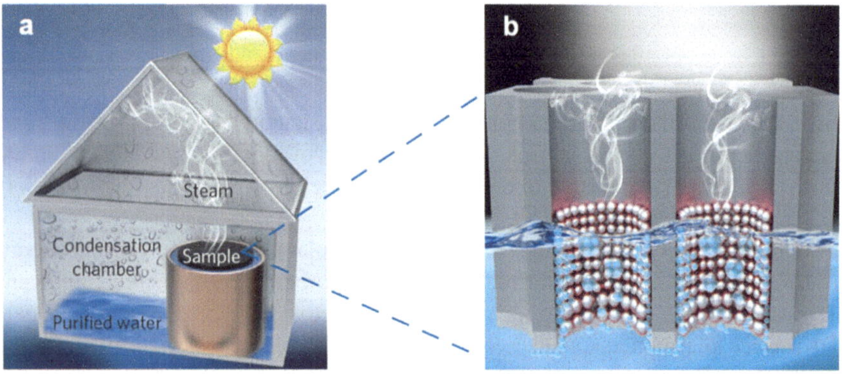


**Fig. 5.12** Solar interfacial desalination: **a** experimental configuration; **b** plasmon-enhanced mechanism. Reproduced with permission form Ref. [219]. Copyright 2016 Nature Publishing Group

the resilience of materials or devices to marine corrosion caused by contaminants such as oil, organic compounds, metal ions, and bacteria. More effort is essential in addressing issues of corrosion and fouling. Consequently, active plasmonic components require surface engineering to maintain stability in harsh environments [206]. Systematically, integrating the desalination process with other industrial activities can synergistically enhance thermal cycles, offering more benefits [208].

### 5.3.3 *Solar Thermal Collector*

A solar collector is a thermal apparatus engineered to convert sunlight into usable heat, predominantly through a black surface that captures solar radiation and transfers the absorbed energy to a proximate working fluid [230, 231]. However, this conversion process is marked by relatively low energy efficiencies, attributed to significant thermal losses via conduction, convection, and radiation, which exacerbate with rising surface temperatures. Volumetric collectors emerge as promising alternatives to augment efficiency, utilizing nanofluids as the working medium (Fig. 5.13) [232–234]. The quantification of useful heat gain is expressed as:

$$Q = C_p \dot{m}(T_{\text{out}} - T_{\text{in}}) \tag{5.3}$$

here, $\dot{m}$ signifies the mass flow rate, and $T_{\text{out}}$ and $T_{\text{in}}$ denote the outlet and inlet temperatures, respectively. Consequently, the thermal efficiency of the collector is calculated based on this expression:

$$\eta = \frac{Q}{IA_c}\% \tag{5.4}$$

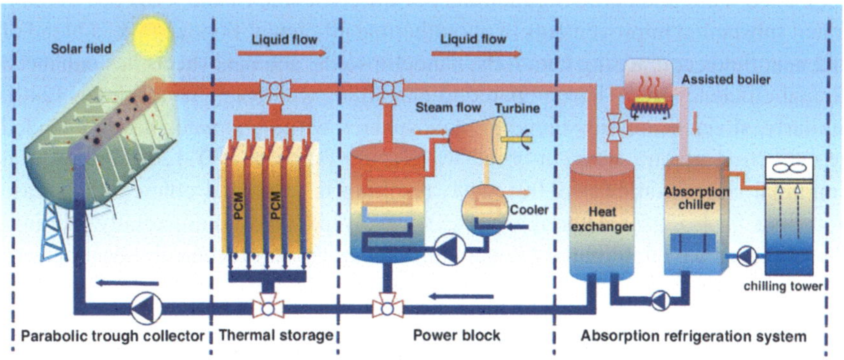


**Fig. 5.13** Thermoplasmonic nanofluid deployed in solar collectors for illustrative purposes. Reproduced with permission form Ref. [241]. Copyright 2019 Elsevier

here $I$ represents the incident radiation density, and $A_c$ denotes the light-harvesting area. For nanofluids, the specific heat and density are specified as: [231]

$$C_{p,nf} = \frac{\rho_{bf} C_{p,bf}(1-\varphi) + \rho_{np} C_{p,np}\varphi}{\rho_{nf}} \tag{5.5}$$

$$\rho_{nf} = \rho_{bf}(1-\varphi) + \rho_{np}\varphi \tag{5.6}$$

where $C_{p,nf}$, $C_{p,np}$, and $C_{p,bf}$ represent the specific heats of the nanofluid, particle, and base fluid, respectively; $\phi$ is the particle volume fraction; and $\rho_{nf}$, $\rho_{np}$, and $\rho_{bf}$ denote the density of nanofluid, particle, and base fluid, correspondingly.

The addition of nanoparticles markedly alters the physical properties of fluids, impacting density, thermal conductivity, specific heat, and viscosity. The low specific heat of the particles facilitates rapid heat absorption compared to pure water. Moreover, their extensive surface area combined with high thermal conductivity promotes the swift transfer of absorbed heat to the surrounding fluid [235], leading to an increase in outlet temperature and enhanced efficiency. Various nanofluids incorporating particles like carbon nanotubes, graphite, noble metals, titanium, and copper oxide have been investigated [233]. Plasmonic nanofluids are particularly appealing due to their strong photothermal properties. Localized surface plasmon resonance significantly enhances sunlight absorption, as the oscillation cross sections of these particles far surpass their physical dimensions [235]. Thus, the collector achieves high thermal efficiency with minimal particle concentrations, rendering these nanofluids exceptionally effective for solar energy harvesting.

Nanofluids containing concentrations of gold or silver below 0.04 wt% have been proven effective for photothermal conversion, though their significant absorption remains restricted to the narrow plasmon band, limiting their solar harvesting capability [236–238]. A more efficacious approach involves combining nanoparticles with varied optical properties to attain broad-band solar absorption [235, 239]. Research on blended nanofluids utilizing materials of different shapes and sizes has demonstrated substantial improvements in photothermal efficiency. For example, a blend of gold nanofluids comprising nanorods, nanoellipsoids, and nanosheets has enhanced thermal efficiency by 104% compared to nanofluids with solely nanospheres [240]. Similarly, silver nanofluids blending nanospheres with nanoplates have expanded the extinction spectra across a broad wavelength range of 300–1200 nm, even at a minimal concentration of 0.001 wt%, resulting in a thermal efficiency increase exceeding 85% [235]. Although not extensively adopted commercially, cutting-edge research has highlighted the economic viability and technical advantages of thermal nanofluids in applications such as energy storage, power generation, and refrigeration, as depicted in Fig. 5.13 [241].

Before practical implementation of plasmonic nanofluids, several challenges must be overcome [232, 233]. The synthesis of these nanofluids through ion-exchange reactions leaves residual ions in the solvent, adversely affecting their application. Introducing a small volume fraction into solar collectors raises concerns regarding material costs and synthesis complexities, detracting from market competitiveness

[242]. Furthermore, the stability of nanofluids poses another challenge; they are prone to agglomeration due to changes in van der Waals forces upon prolonged solar exposure. While integrating plasmonic nanostructures into the absorbing layer of solar collectors can improve absorption efficiency over a broad solar spectrum [243], the cost-effectiveness and scalability of these nanostructures and their fabrication processes are critical for commercial viability.

### 5.3.4 Radiative Cooling/Energy-Saving Window and Anti-icing/Fogging

The assembly of metallic surfaces composed of individual plasmonic structures enables the absorption or scattering of light at nanoscale. Using various dielectric materials in metasurfaces opens new avenues for applications such as radiative cooling, energy-efficient windows, and innovative methods for preventing fogging or icing [244–246]. The localized surface plasmon resonances of nanostructures dictate their optical behaviors, resulting in either light absorption or reflection. However, an inherent limitation arises from the spectral overlap between scattering and absorption, attributed to the material properties of the components [247]. To enhance functionality for the applications, advancing the design of optical and thermal structures is essential for fine-tuning electromagnetic flux and achieving comprehensive control over the absorption and scattering spectra. The introduction of a hybrid metal–dielectric antenna concept allows for precise manipulation of these processes [248]. Adjustments to the system's geometry enable flexible modulation of the scattering-to-absorption ratio, along with their relative enhancement or suppression across a wide spectrum. Consequently, by integrating free-space radiation with a hyperbolic antenna, regions of radiative and nonradiative phenomena can be strategically designed, ensuring managed interaction across both channels (Fig. 5.14).

Radiative cooling involves expelling heat from surfaces into distant space via infrared radiation, often by reflecting solar radiation across a broad or selective spectrum [245, 249]. Fine-tuning the radiative properties is essential for creating selective radiators with distinct infrared characteristics. For instance, a scalable plasmonic thin film has been developed by embedding randomly dispersed dielectric microspheres into a polymeric matrix for sub-ambient cooling. This metamaterial film remains fully transparent to visible light while exhibiting an infrared emissivity exceeding 0.93 in the atmospheric window, achieving a radiative cooling power of 93 $W/m^2$ under solar irradiation when back-coated with silver [250]. Additionally, a Janus emitter, comprising Ag-polydimethylsiloxane on a quartz substrate, has been designed to minimize emissivity loss [251]. This structure successfully separates selective and broadband emissions on its two sides, facilitating substantial surface cooling and passive reduction of the greenhouse effect within enclosures.

Plasmonic surfaces consist of arrays of basic optical elements that foster specific plasmon resonances and demonstrate a tunable, directionally and spectrally selective

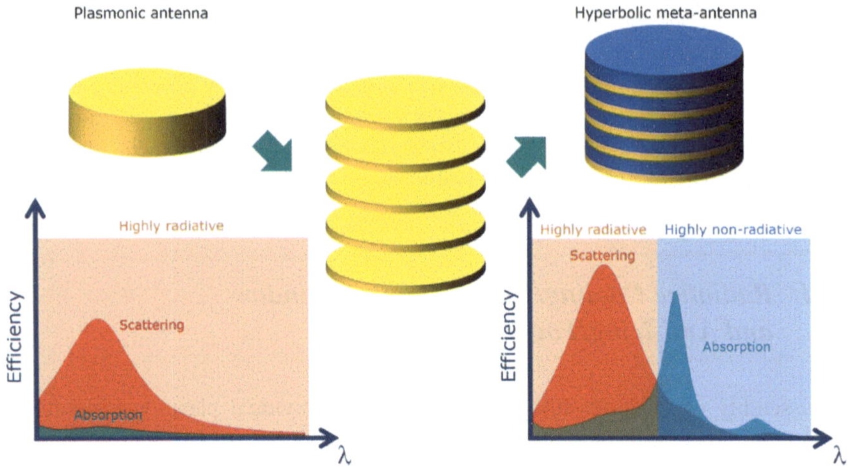


**Fig. 5.14** Plasmonic antenna (left panel) exhibits a pronounced radiative spectral response, while the incorporation of a dielectric spacer between antenna slices (right panel) facilitates distinct radiative and nonradiative regions. Reproduced with permission form Ref. [248]. Copyright 2019 American Chemical Society

light absorption, capable of generating heat through NIR absorption while maintaining transparency in the visible spectrum [247]. Such surfaces offer innovative approaches for designing energy-efficient windows with optimized energy transfer in buildings. For instance, an energy-saving window featuring nanoarrays of multielement gold and nickel plasmonic antennas on a glass substrate elevates the window's glazing temperature by up to 8 °C under solar radiation without altering visible-light transmission or color. Similarly, a gold particle-chain solar radiator can increase the glazing temperature by as much as 9 °C, while still allowing high visibility [252]. These advancements present new opportunities for developing smart windows that incorporate transparent heating elements powered by sunlight.

Surface icing or fogging presents challenges for wind turbines, airplanes, and power lines. The current remedies rely on mechanical or chemical removal, or on-site electrical heating strategies that are energy-intensive, costly, and technically demanding [253, 254]. Recent developments have focused on solar anti-icing/frosting surfaces that harness sunlight to convert light energy into heat, thereby inhibiting ice formation [244, 255–257]. Plasmonic surfaces intensify sunlight at small volumes, substantially increasing the temperature at the air/material interface where condensate or icing is prone to occur, thus deterring freezing or reducing ice adhesion. The efficacy of these surfaces hinges on the maximum temperature rise and thermal response under solar radiation [254, 258]. By engineering plasonomic structures, heat can be efficiently spread laterally across the interface while maintaining transparency. Even under super-cooled conditions, this transparent metasurface, integrated with hierarchical micro/nano textures, effectively prevents frosting and water nucleation [259].

However, challenges exist in making this technology viable for large-scale or practical use that includes the cost of fabricating plasmonic materials, the durability, and longevity of the nanocoatings under various environmental conditions. Advances in material science and plasmonic research continue to push the boundaries of what is achievable in creating efficient and reliable solutions.

### 5.3.5 Thermoplasmonic Motor

A thermoplasmonic motor is a nanoscale device that converts light energy into mechanical energy via thermoplasmonic effects. This motor utilizes localized heating from plasmonic nanostructures upon light absorption, primarily from laser sources. The mechanisms for motion generation include [260, 261]: (1)Thermally induced surface tension variations along fluid interfaces, known as the Marangoni effect, propel microscale or nanoscale devices through specifically designed thermoplasmonic structures [262, 263]. (2) Photons possess mechanical properties, exerting momentum transfer or optical forces on objects through mechanisms of scattering or absorption [264]. (3) Nanomotors embedded with plasmonic photocatalysts transform energy from both light and chemical fuels into mechanical movement [265, 266]. (4) Plasmonic heating may also trigger phase transitions, such as from liquid to gas, generating bubbles whose rapid expansion or collapse produces propulsive forces [267].

In thermoplasmonic motors, nanoparticles are embedded in or coated on substrates that are responsive to thermal fluctuations [268, 269]. Upon irradiation, these nanoparticles generate surface temperature gradients, driving thermocapillary flows or thermophoretic movements that can propel microscale or nanoscale entities [270]. A typical design features Janus $SiO_2$ nanomotors with a sputtered 10 nm gold layer on one hemisphere. Here, a near-infrared laser locally heats this Au segment, establishing a thermal gradient that induces opto-thermophoretic propulsion against the orientation of the Au-coated side, achieving speeds up to 950 body lengths per second at a laser intensity of 703 $kW/m^2$ [271]. Additionally, rotational forces are generated through photon momentum [272–275]. By altering the light wavelength to excite distinct plasmonic modes, these motors can rotate silica microdisks significantly larger in volume by four orders of magnitude [264]. Concurrently, photocatalytic activities on the particle surfaces may produce minuscule gas bubbles; expelling these from the particle in a directed manner complies with momentum conservation principles, propelling the particle inversely [265]. Furthermore, nanobubble generation via photothermal activation of absorptive particles has been achieved using wavelength-specific illumination targeted at particle localized surface plasmon resonance peaks with both continuous and pulsed light [276–278]. Employing this method, miniature boat-like structures with plasmon-active surfaces navigate across liquids driven by the expulsion of hot vapor jets, presenting novel methodologies for directing the movement of photothermal motors (Fig. 5.15) [267].

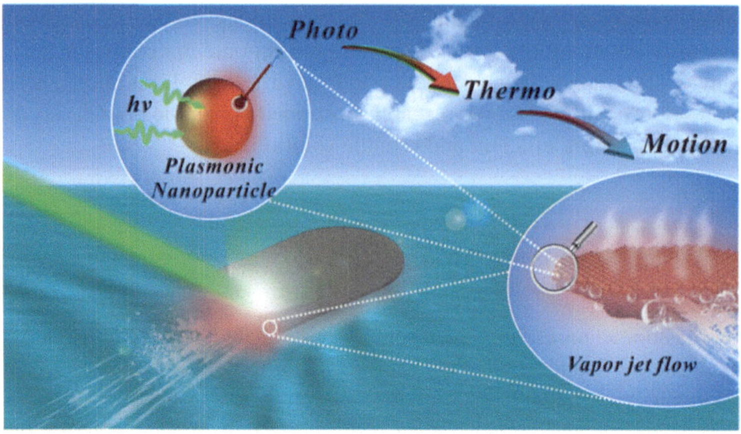


**Fig. 5.15** Thermoplasmonic motor propels forward at the liquid/air interface by ejection of a vapor jet. Reproduced with permission from Ref. [267]. Copyright 2017 American Chemical Society

The benefit of thermoplasmonic motors lies in their light-driven, wireless operability, enabling distant activation and manipulation [261]. Their precision control at micro and nanoscopic levels renders them suitable for diverse applications such as targeted drug delivery, microfluidics, sensing, and nanostructure assembly [279]. The development of efficient, reliable thermoplasmonic motors necessitates design optimization to enhance energy conversion, ensure directional movement control, and assure sensitivity to specific light wavelengths. Progress in nanomanufacturing technologies, including electron beam lithography and colloidal synthesis, is imperative for fabricating the intricate nanostructures these motors require [280, 281]. Despite their promise, thermoplasmonic motors face challenges like managing thermal dissipation to avoid system damage, minimizing energy losses, and achieving scalable production without compromising nanoscale features.

## 5.4 Chemical Science

### *5.4.1 Chemical Synthesis*

Chemical reactions play pivotal roles in daily life, where catalytic processes in chemical synthesis can be conducted either thermally or through photocatalysts [282]. However, thermal catalysis lacks control over surface selectivity in chemical transformations, relying on the manipulation of activation barriers and adsorption energies. Thermoplasmonics, by confining heat close to thermodynamic limits, offers new potential in chemical synthesis via controlled nanoscale thermal environments [283, 284]. This method enhances local heating, excites hot electrons and amplifies the magnetic field near the surface, significantly promoting chemical reactions [285–287]. Efficient energy transfer between metal surfaces and adsorbates initiates chemical transformations in accordance with the Arrhenius law [288, 289]. The synergistic optimization of photoelectron-thermochemical contributions in solar-driven plasmonic catalysis can substantially improve selectivity and reaction rates, even under ambient conditions [290, 291]. Consequently, thermoplasmonic catalysis serves as a crucial complement to traditional methods in chemical synthesis [292].

Plasmonic nanostructures function as photocatalysts in either direct or indirect photocatalysis. In the direct approach, the surface of plasmonic nanostructures serves as catalytic active sites that absorb light and facilitate chemical transformations upon resonance excitation [293]. Recent strategies have involved integrating plasmonic metals with traditional photocatalysts/$TiO_2$, to boost the activity and harness the full solar spectrum [294]. Photocatalysis on the surface of metallic particles demonstrates a heterogeneous mechanism. In contrast, a homogeneous mechanism is observed when it occurs in solution with ions leached from the nanoparticle surfaces [295]. Furthermore, plasmonic heating has been utilized for chemical reactions, such as catalytic ethanol reforming in a microfluidic reactor containing Au nanoparticles [296]. Upon laser exposure at the plasmon frequency to an ethanol–water mixture, gas products including CO, $CO_2$, and hydrogen emerge from endothermic reactions driven by plasmonic heating. Additionally, plasmonic Au–Pd and $Cu_7S_4$@Pd nanostructures have been reported for enhanced Suzuki reactions, achieving reaction yields twice those of thermal reactions at the same temperature under solar radiation [297, 298]. Moreover, thermoplasmonic catalysis employing Au and Ag nanocrystals, Pd nanostructures, and TiN cylindrical nanocavities has also been effective in accelerating processes such as CO oxidation, hydrogenation of olefins, and ethylene epoxidation [299–302].

Enhancing reaction selectivity and efficiency are main objects in chemical synthesis, yet they often exhibit reciprocal competition [303, 304]. Recent approaches utilize surface plasmon resonance to modulate selectivity, tackling this dual challeng [304–307]. Tian et al. demonstrated that SPR excitation fulfills both goals, yielding high selectivity and efficiency in the synthesis of acrolein or propylene oxide. The mechanisms involved in selective propylene oxidation are depicted in Fig. 5.16, where hot electrons promote oxygen activation, steering the reaction selectively toward acrolein. Meanwhile, the nanoconfined SPR effect localizes the active region, preventing overoxidation and thus enhancing selectivity [307]. However, our understanding of the dual impacts on selectivity and efficiency remains nascent, particularly in nanoconfined reactions that engage both hot carriers and plasmon heating [308, 309]. Such scenarios, characterized by inhomogeneous fields including enhanced magnetic, thermal, or chemical gradients, can significantly alter the reaction. Additionally, the tendency of plasmonic materials used as photocatalysts to agglomerate poses a disadvantage by diminishing photocatalytic activity. As research advances, the development of innovative synthetic methodologies leveraging the unique properties of plasmons is anticipated to foster sustainable chemical manufacturing.

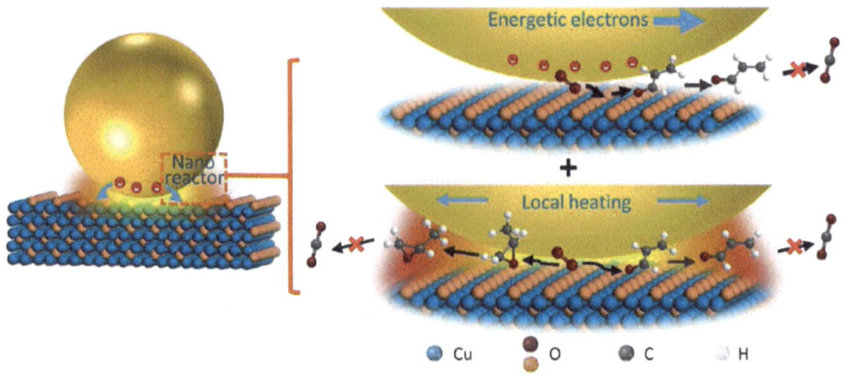


**Fig. 5.16** Solar chemical synthesis involves both energetic electrons and localized heating effects driving the chemical reaction of propylene over Au-$Cu_2O$ catalysts. Reproduced with permission form Ref. [307]. Copyright 2021 AAAS

## 5.4.2 *Solar Hydrogen*

Using solar energy for sustainable hydrogen production represents a promising avenue in energy sector [310, 311]. The cornerstone of solar-to-hydrogen conversion lies in the synergistic design of photothermochemical catalytic processes to booster the water-splitting reaction. By optimizing light absorption and energy confinement, plasmonic catalysts exhibit great potential to enhance this reaction through photocatalytic and/or photoelectrochemical means [312–315]. This enhancement mainly operates via three pathways: hot-carrier generation, electromagnetic field amplification [316], and the plasmon heating effect [288, 317–319]. Given that redox reactions predominate, efficient charge transfer, critical for photocatalysis, hinges on the separation and transport of energetic carriers to drive these reactions [320, 321]. Moreover, the operation temperature influenced by the Nernst equation impacts redox reactions. Thus, the plasmonic heating effect, even if not the primary mechanism, must be judiciously managed in solar water splitting. In this paradigm, sunlight is harnessed to produce hydrogen gas from water through photochemical or photoelectrochemical pathways, with plasmonic nanoparticles enhancing efficiency via plasmonic heating.

Solar water oxidation utilizing Au/$TiO_2$ photocatalysts reveals that hot holes localized near the heterojunction interface are crucial for water oxidation. Density functional theory underscores the role of the interfacial Ti–O–Au structure in advancing oxygen evolution, lending further credence to these observations [322]. Au–$TiO_2$ Janus catalysts, featuring confined plasmon near-fields at the heterointerface, prove beneficial for visible-light-driven hydrogen production [323]. Localized electronic states in $TiO_2$, strongly coupled with near-field interactions at the interface, significantly boost light absorption and electron–hole pair generation, thus enhancing the reaction. Contrary to individual plasmonic structures, the collective excitations in embedded metallic particles and composite materials emerge as a more effective approach to augment solar water splitting efficiencies [324–328]. For instance, hydrogen evolution from Au-nanochains increases by 3.5 times compared to conventional isolated particles due to the intensified electromagnetic fields generated by the coupled plasmonic structures [329]. Integration of plasmonic materials can particularly improve photoelectrochemical cells, which often suffer performance limitations due to discrepancies between semiconductor electrodes' optical absorption properties and the solar spectrum. Further enhancement in water splitting is achieved using a gold nanoparticles/$TiO_2$/Au-film sandwich structure as the photoelectrode; this configuration facilitates robust modal couplings between the F-P modes of the hybrid film and the LSPR of gold particles [326]. This arrangement not only captures a broader spectrum of solar radiation but also concentrates energy at the active sites, thereby minimizing energy losses and accelerating reaction rates.

Thermoplasmonic water splitting, akin to thermal catalysis with external heating [288], yet the assessment of thermal contribution to the reaction rate remains under debate [330–333]. Li et al. first report of hydrogen production using $TiSi_2$ and Pt/$TiO_2$ catalysts under visible light, highlighting an essential pressure increase for hydrogen generation at 190 °C [334]. To advance seawater catalysis, core–shell $SiO_2$/Ag@$TiO_2$ nanostructures have been developed, leveraging their superior photothermic properties as depicted in Fig. 5.17. A solar–thermal reactor, comprising a quartz tube and a parabolic trough concentrator, facilitates a high energy flux, wherein Ag-induced plasmonic heating generates a high-temperature boundary at the interface, initiating both seawater catalysis and steam production processes. Field tests confirm this setup's enhanced performance in dehydrogenating oxygenates, with the core–shell configuration yielding a significant ~ 95% increase (1536 mmol $g^{-1}$ $h^{-1}$) in the hydrogen generation rate (Fig. 5.17b) [216]. This evidence underscores the potential of thermoplasmonic hydrogen production under natural sunlight for efficient solar-driven seawater catalysis, producing both hydrogen and freshwater.

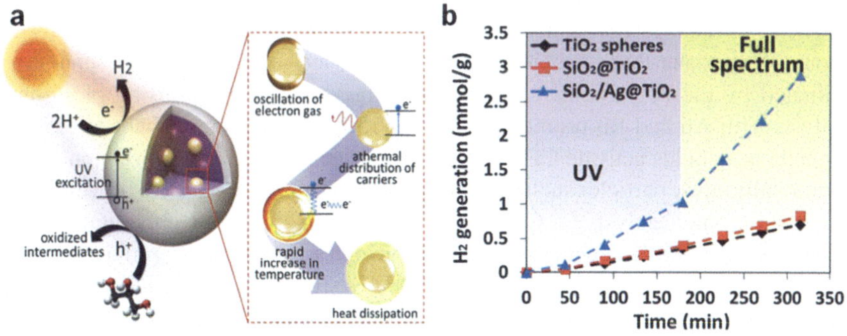


**Fig. 5.17** Solar hydrogen generation **a** a schematic depiction of the mechanism, **b** hydrogen production using $TiO_2$ spheres and $SiO_2$@$TiO_2$, $SiO_2$/Ag@$TiO_2$ core–shell nanocomposites in a glycerol–water solution under UV and full-spectrum light irradiation. Reproduced with permission from Ref. [216]. Copyright 2016 Royal Society of Chemistry

Thermoplasmonics offer an innovative strategy for enhancing solar hydrogen production by enhancing light absorption and increasing local temperatures, bringing us closer to produce sustainable hydrogen with the abundant solar energy [330, 335]. Despite these advantages, challenges remain in utilizing plasmonic heating for hydrogen production. Durability under solar illumination, optimizing the particle size and shape for maximum plasmonic effect, ensuring efficient hot electron transfer, and the high cost of noble metals are issues that need to be resolved to make this approach economically viable.

### 5.4.3 Solar $CO_2$ Reduction

The conversion of carbon dioxide into high-value chemicals is crucial for the sustainable transformation of the chemical industry and combating climate change [336, 337]. Solar-driven $CO_2$ reduction emerges as a highly promising approach, storing renewable energy in hydrocarbon forms while mitigating $CO_2$ emissions [338]. Since Halmann's initial discovery of $CO_2$ reduction into organic compounds [339], there has been a global surge in research on solar-to-fuel conversion, though this field remains largely at the proof-of-concept stage [340–343]. Challenges include optimizing light absorption, activating $CO_2$ molecules, and facilitating charge generation and transport for $CO_2$ reduction into hydrocarbons. This complex process necessitates the synergistic interaction between electrons and active sites on catalysts to efficiently promote C–H bond formation while minimizing side reactions. Recent developments highlighted the role of metallic nanostructures with localized surface plasmon resonance in enhancing catalytic efficiency [344–346].

$CO_2$ reduction over plasmonic silver catalysts has been investigated at single-particle spatial resolution, highlighting the critical role of photogenerated HOCO* intermediates from surface/adsorbate and photoexcited state interactions in efficient solar fuel production [347]. $TiO_2$ crystals coated with ultrafine Pt particles have demonstrated exceptionally high photoreduction efficiency for $CO_2$, selectively producing methane due to rapid carrier transport and effective charge separation facilitated by the Pt particles [348]. Thermoplasmonic catalysis using plasmonic catalysts with Au and Rh promoters on SAB-15 has shown that enhanced electromagnetic near-fields activate $CH_4$ and $CO_2$ for selective methane conversion [349]. Additionally, alloy particles such as Au-Cu [343, 350], and Cu-Ru have proven effective as plasmonic catalysts for $CO_2$ reduction, where Au and/or Cu facilitate strong light absorption and efficient carrier generation, and single-atom Ru provides highly active reaction sites [306]. Furthermore, rational design of plasmonic heterostructure photocatalysts, such as metal/insulator/semiconductor configurations or those with polar facets, can achieve highly selective $CO_2$ reduction [351, 352].

Compared to plasmonic catalysis, thermoplasmonic catalysis utilizes the full spectrum of sunlight by converting photons into heat [353, 354], enhancing conversion performance through the heating effect [342, 355–357]. The thermoplasmonic hydrogenation of $CO_2$ has been explored using Pd@$Nb_2O_5$ nanorod catalysts, where localized temperatures induced by Pd nanoheaters reach up to 470 °C, sufficient to drive the endothermic reactions. These conditions enable small Pd nanocrystals to achieve a CO selectivity of 99.5% with a notable conversion rate of 18.8 mol $h^{-1}g_{Pd}^{-1}$ [356]. Inspiration from the greenhouse effect, recent developments include supra-photothermal catalysts made of nanoporous-silica-encapsulated nickel nanocrystals (Ni@p-$SiO_2$), which generate local temperatures significantly higher than those of conventional nickel catalysts to enhance $CO_2$ hydrogenation (Fig. 5.18) [357]. This enhanced performance is primarily due to the supra-photothermal effect at the nickel core (Fig. 5.18b), bolstered by infrared shell shielding and thermal insulation, effectively confining the energy.

The conversion of $CO_2$ into value-added chemicals such as carbon monoxide, methane, methanol, and other hydrocarbons is pivotal for sustainable energy initiatives. Photothermal $CO_2$ reduction in plasmonic catalytic systems integrates multi-physical processes across different timescales [358, 359]. The concurrent photochemical and thermochemical processes make thermoplasmonic catalysis a powerful approach for heterogeneous catalysis in solar energy conversion [354]. Early attempts demonstrated the feasibility of using solar energy to transform $CO_2$ and water into

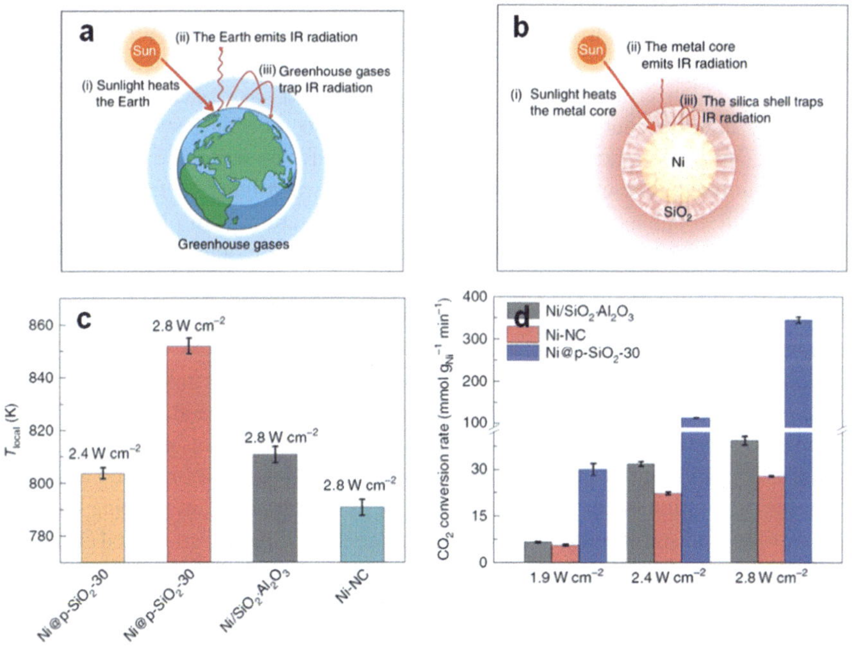


**Fig. 5.18** Solar $CO_2$ reduction (**a**, **b**) Nanoscale greenhouse effect; **c** Predicted local temperature; and **d** $CO_2$ conversion rate. Reproduced with permission from Ref. [357]. Copyright 2021 Nature Publishing Group

fuels [350]. However, challenges persist in the mechanistic understanding, selective product control, and overall low efficiencies [336, 340]. Moreover, practical issues such as optimizing nanoparticle size, shape, and composition for maximal light absorption and heat generation, enhancing catalyst stability during continuous operation, and improving the integration of plasmonic components with catalytic materials for efficient energy transfer and reactant conversion must be addressed.

### *5.4.4 Nitrogen Fixation*

Fertilizers are essential for enhancing agricultural crop yields. The Haber–Bosch synthesis of ammonia from dinitrogen, a significant discovery of the twentieth century, has greatly augmented human survival by enabling synthetic fertilizer production and has profound implications in basic science [360–363]. Nonetheless, this process is energy-intensive, consuming 1–3% of global annual energy and necessitating high temperatures and pressures, which reduce equilibrium conversion and increase capital costs [364]. These limitations have spurred research into alternative $N_2$ fixation methods. Solar thermocatalytic conversion of atmospheric dinitrogen into nitrogen products offers a sustainable means to produce solar fertilizers, minimizing the energy and carbon footprint associated with traditional ammonia synthesis and facilitating integration with existing infrastructures [360, 365, 366]. Challenges remain, including the selection of effective photocatalysts and system integration for enhanced solar fertilizer production. Recently, LSPR has gained prominence for its efficacy in solar energy conversion attributable to its light-harvesting capabilities [367, 368]. Metallic structures leveraging LSPR show significant promise in boosting photocatalytic efficiency due to mechanisms involving $N_2$ fixation through LSPR near-field enhancement, direct hot electron injection from plasmonic metals into $N_2$, and hot electron transfer to semiconductors among others [359].

A sandwich-like nanostructured electrode ($TiO_2$/Au/a-$TiO_2$) has been developed to enhance nitrogen fixation. The presence of surface oxygen vacancies in the outer amorphous $TiO_2$ layer improves nitrogen adsorption and activation, facilitating ammonia synthesis through hot electron transfer from plasmonic gold particles to the semiconductor $TiO_2$ [369]. For improved light harvesting, Au and Ru particles are separately loaded on opposite sides of an Nb-doped $SrTiO_3$ film, serving as cocatalysts. This configuration, when co-loaded with Au and Ru, demonstrated effective ammonia production under solar radiation, with its action spectrum aligning with the gold particles' plasmon band [370]. Similarly, a gold and Zr/$ZrO_x$ co-loaded electrode has been proposed for ammonia synthesis via plasmon-mediated charge separation, further emphasizing the critical role of plasmon resonance in enhancing solar fertilizer production [371]. Additionally, integrating this system with a photovoltaic cell presents a viable approach to boost efficiency. A solar-driven photoelectrochemical cell based on plasmon-enhanced black silicon has achieved nitrogen conversion to ammonia, yielding a high rate of 13.3 mg $m^{-2}$ $h^{-1}$ under two suns [372]. This superior performance is attributed to robust light absorption, carrier generation, effective charge separation, and the synergistic effect of additional cocatalysts.

The yield of the ammonia synthesis reaction is constrained by equilibrium laws due to the balance between nitrogen dissociation and ammonia decomposition at high operational temperatures [373, 374]. To overcome the limitations, a solar–thermal mechanism utilizing $K/Ru/TiO_{2-x}H_x$ catalysts has been designed for enhancing ammonia synthesis (Fig. 5.19) [375]. This photothermal catalyst exhibits significant sunlight absorption, raising its temperature to 360 °C through effective thermal conversion. The Ru catalyst, embellished on the surface, facilitates nitrogen activation at ambient conditions via hot electron transfer from $TiO_{2-x}H_x$. In contrast to traditional thermal catalysis, this heterostructure effectively confines electromagnetic fields and heat flux around the Ru clusters, considerably lowering the activation energy and increasing catalytic activity (Fig. 5.19b). Further innovation includes thermoplasmonic catalysis using a Fe-nanonecklace hydrogenated titanium hybrid, which introduces a dual-temperature-zone feature to enhance performance [374]. This configuration achieves a surface temperature of 495 °C with a significant temperature gradient of 137 °C between the hot (Fe) and cold ($TiO_{2-x}H_y$) zones. The hot zone generates hot electrons that efficiently dissociate nitrogen, while the cold zone supports the sequential hydrogenation of migrated nitrogen atoms from the Fe domain, markedly suppressing the reverse reaction and enabling yields that surpass conventional equilibrium limits. These advancements offer promising avenues for solar-driven ammonia synthesis.

By localizing heat with plasmonic nanoparticles in solar fertilizer, it is possible to utilize solar energy to drive the nitrogen fixation reaction at much lower overall temperatures and pressures [376–378]. Such solar fertilizer production aligns with the prospects of green ammonia. While the technology is in its nascent stages and faces challenges, including scalability, cost-effectiveness, and stability of the plasmonic materials [361], ongoing research and advances in nanotechnology and materials

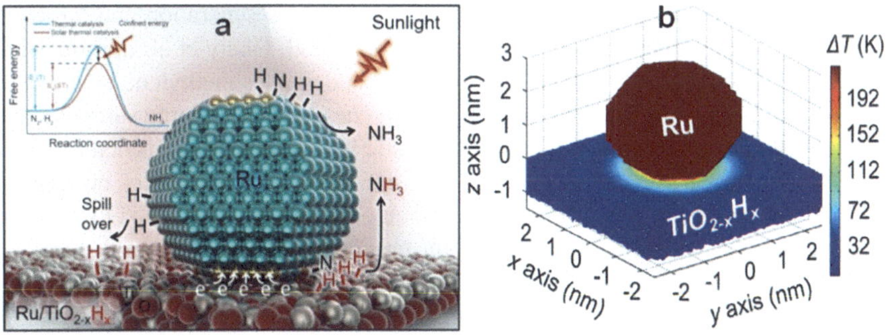


**Fig. 5.19** **a** Solar ammonia synthesis. **b** Temperature profile around Ru cluster. Reproduced with permission from Ref. [375]. Copyright 2018 Elsevier

science promise innovative solutions to overcome these barriers, eventually leading to a sustainable approach to chemical manufacturing.

## 5.5 Environmental Science

### 5.5.1 Water Purification

As industrialization expands, organic pollutants, particularly organic dyes from chemical and textile sectors, increasingly threat our ecosystems [379]. These dyes, characterized by their complex structures and stability, prove highly resistant to biodegradation and pose significant toxicological and carcinogenic risks. Consequently, their removal from aquatic environments becomes imperative [282, 380]. One sustainable method for addressing this challenge is solar photocatalytic degradation, a green technique in environmental purification [338, 381, 382]. This method has spurred the development of various photocatalysts, among which metal/semiconductor plasmonic catalysts emerge as particularly effective [383, 384]. The superiority of these plasmonic catalysts primarily rests on their enhanced visible-light absorption capabilities attributable to the surface plasmon resonance effect. Additionally, these catalysts facilitate efficient carrier separation in the semiconductor's conduction band via direct charge injection and plasmon-induced field enhancement [385], significantly boosting photocatalytic performance.

The efficacy of plasmonic structures in water purification has been established, wherein their pronounced local field enhancement facilitates on-site pollution detection via Raman scattering and promotes advanced pollution remediation [386]. Notable among these is bifunctional Au@$TiO_2$ nanoparticles [387], which concurrently produce clean water and degrade pollutants through photocatalytic activity [388]. Similarly, Au-embedded $TiO_2$ nanofiber catalysts demonstrate potent photodegradation of organic dyes such as Rhodamine B and methylene blue, with improved performance attributed to effective charge separation and targeted visible-light absorption within the resonance band [389]. Ghaly et al. introduced a stable plasmonic Ag/AgCl-polyaniline catalyst that degrades methylene blue under sunlight, exhibiting superior degradation effectiveness due to its large surface area, strong visible-light harvesting, and the synergistic interaction within its heterostructure [390]. Additionally, Au particles of varied sizes and morphologies on $Bi_2O_3$/$Bi_2O_2CO_3$ structures significantly enhance photocatalytic activity by 4–11 times for dye degradation, with activity enhancement particularly observed across different types of gold nanorods. This high activity is due to the effective charge separation with more production of ·OH radicals from the plasmonic structures [391]. Thus, the catalytic efficiency is substantially influenced by the structural and morphological characteristics of the plasmon, the substrate's material properties, and the interactive dynamics between the plasmonic components and the substrate, emphasizing the critical role of structural engineering in optimizing photocatalytic performance.

Plasmonic particles exhibit notable photocatalytic capabilities; however, their hot electrons are excited within restricted wavelengths. If not utilized in charge transfer, these energetic carriers may increase particle temperature, a phenomenon that can be redirected to enhance catalysis at active sites through localized heating [386, 392]. To differentiate thermal from nonthermal effects, plasmonic photocatalysis has been evaluated under controlled illumination conditions [342, 393, 394]. For instance, gold nanorods partially coated with silica have demonstrated high activity in the oxidative degradation of complex organic micropollutants such as bisphenol A and amoxicillin under both dark and illuminated conditions (Fig. 5.20). The contributions to enhanced oxidation include direct electron transfer in darkness, photothermal heating, and hot electron transfer under light exposure [395]. Additionally, systems like Pt/$\gamma$-$Al_2O_3$ have been explored for the decomposition of toluene under solar radiation, leveraging the thermoplasmonic effect of Pt particles that serve dual roles as solar absorbers and active catalytic sites [396]. These investigations pave the way for thermoplasmonic strategies in solar-driven remediation, offering a promising avenue for addressing persistent water pollutants.

Thermoplasmonics presents a promising, innovative approach for sustainable water purification. Its principal benefit resides in the selective heating of target contaminants, offering avenues for real-time process monitoring and control. By observing plasmonic responses, operators can dynamically fine-tune operational parameters to maximize treatment efficacy while accommodating fluctuating water qualities [397]. However, the progression from laboratory-scale prototypes to full-scale implementation poses significant challenges. The introduction of novel technologies, particularly those impacting vital resources like drinking water, entails comprehensive evaluation and regulatory certification [398], highlighting the complexity of transitioning thermoplasmonic solutions to widespread practical use.

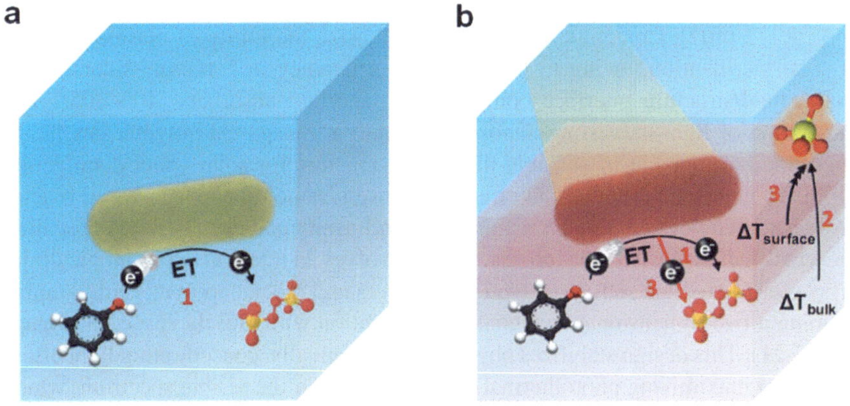


**Fig. 5.20** Phenol degradation in water, and the mechanisms under **a** dark or **b** solar irradiation. Reproduced with permission [395]. Copyright 2020 National Academy of Sciences

### 5.5.2 *Sterilization and Disinfection*

Inadequate access to potable water represents a critical challenge globally, particularly in rural locales. Solar disinfection emerges as a promising solution in areas with substantial sunlight but limited infrastructure [398]. Nevertheless, the optimal and most reliable solar disinfection method, capable of efficiently utilizing solar energy, remains to be determined despite technological advancements. Beyond mitigating airborne diseases, sterilization is crucial for controlling waterborne pathogens [399, 400]. Bacterial viability declines at temperatures exceeding 45 °C, endorsing steam exposure as a prevalent sterilization method. The efficacy of steam sterilization hinges on both the steam's temperature and exposure duration. Sterilization mechanisms involve mechanical damage and thermal denaturation [401], the latter occurring prior to reaching the critical temperature and primarily involve protein denaturation through direct impacts on ribosomes, enzyme structure disruption, and cellular inactivation. This thermal damage is effective above a specific denaturation temperature $T_{\mathrm{D}}$, with the duration of thermal denaturation $\tau_D$ relative to the operating temperature $T$ [401].

$$\tau_D = \frac{\hbar}{k_B}\exp\left(\frac{\Delta H - T\Delta S}{R_g T}\right) \tag{5.7}$$

whereas the thermomechanical damage occurs through the emission of shock or pressure waves during explosive boiling, disrupting cell structures and leading to bacterial destruction.

Thermoplasmonics, initially applied for sterlization of medical instruments using concentrated solar radiation on gold nanoshells to produce steam, demonstrate efficacy in bacterial and viral inactivation when nanofluids containing Au nanorods and carbon blacks are employed, achieving significant log reductions in E. coli, MS2, and PR772 [402]. Contemporary studies emphasize the utility of plasmonic films in solar sterilization, owing to their superior efficiency in confining solar energy at the fluid/structure interface compared to nanofluid absorbers [403–405]. The integration of light-harvesting nanoparticles into a transparent polymer has facilitated the development of scalable film-based reactors for solar water disinfection [406]. To enhance photothermal effects, a dense plasmonic nanogap-based reactor array has been devised, though maintaining uniformity and density while upscaling presents challenges. An innovative solution involves an integrative plasmonic dimer array constructed via self-assembled monolayers and block copolymer lithography, forming a two-dimensional hexagonal configuration with closely spaced nanogaps (Fig. 5.21). This design stabilizes physically and optically, concentrating light effectively and maximizing photothermal energy capture in the visible spectrum, which markedly improves E. coli disinfection efficacy up to fivefold [407].

A novel near-field radiative reactor utilizing concentrated heat strategies has been further suggested to enhance thermal flux [408]. This device features a spatial separation between the sterilization and absorption surfaces, effectively addressing issues

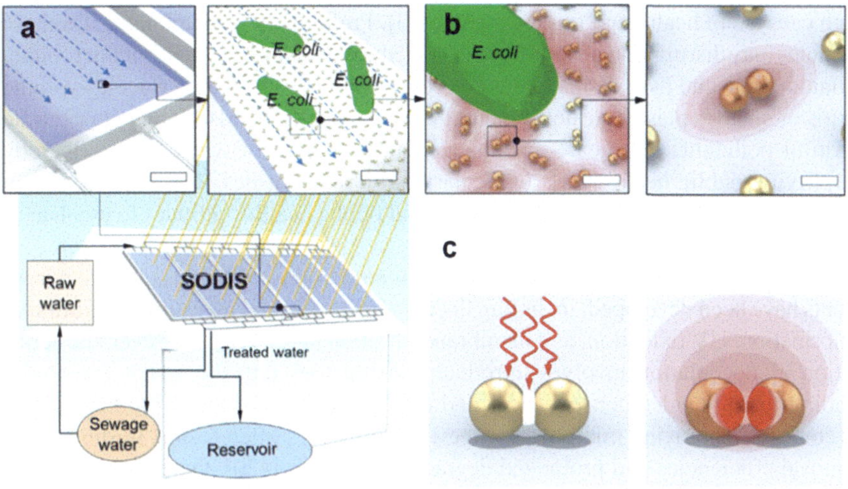


**Fig. 5.21** Solar water disinfection **a** integrating optofluidic panels, outfitted with a uniform and high-density plasmonic dimer array, onto rooftops; **b** utilizing the thermal energy generated by the dimer array to inactivate waterborne pathogens; **c** leveraging the focused electromagnetic field within the nanoreactor to enhance the thermoplasmonic effect. Reproduced with permission form Ref. [407]. Copyright 2023 American Chemical Society

of pore blockage and surface contamination. Ongoing research explores the optimal deployment of plasmonic nanoparticles for disinfection applications. Enhanced treatment efficacy may be achieved through the integration of plasmonic sterilization with bubble technologies, such as nucleation or cavitation [401]. The generation of small bubbles not only boosts biological water treatment but also effectively inactivates pathogens and reduces biofouling. Additionally, the mechanical, thermal, and chemical effects induced by cavitation contribute to the inactivation of microorganisms. If realized, this technology could establish new sterilization protocols and devices that offer clean, efficient alternatives to conventional methods, particularly suitable for contexts where chemical treatments or high temperatures are undesirable.

### *5.5.3 Removal of Nano/Microplastics*

Public concern is mounting over the contamination from plastic debris, intensified by the accelerated production of plastics [409, 410] and exacerbated by their prolonged degradation rates and inadequate recycling efficiencies [411]. Through processes such as mechanical abrasion, UV radiation exposure from sunlight, or biological degradation, these plastics break down into microplastics (1 μm–5 mm) and potentially even smaller nanoplastics (< 1 μm), which are readily transported into aquatic systems [412]. These minuscule plastic fragments can permeate the food chain and accumulate in higher trophic levels, leading to uncertain implications for both ecosystem health and food security [413]. Preliminary research indicates these particles may disrupt iron absorption and contribute to disturbances in gut microbiota, inflammation, and liver damage [414, 415]. Additionally, the vast surface-to-volume ratios of smaller plastic particles facilitate their capacity to adsorb and concentrate harmful pollutants like heavy metal ions and pesticides in water bodies, thereby amplifying public health risks, particularly with nanoplastics. The impact of such nano/microplastics on aquatic ecosystems and public health continues to escalate as a focal point of concern [416].

Current techniques for the extraction of nano/microplastics from aquatic environments have been developed, including flocculation, microbial ingestion, biofiltration, chlorination, UV oxidation, and membrane filtration [417, 418]. However, the application of flocculation involves introducing metal ions into the water, potentially leading to unintended consequences and the generation of toxic byproducts [419]. In contrast, employing microbes to ingest microplastics often results in suboptimal removal efficiencies and prolonged degradation periods [420]. Oxidation technologies might increase the toxicity and elevate the presence of hazardous contaminants, owing to changes in surface topography and chemical properties caused by incomplete oxidation processes. While highly efficient and devoid of secondary pollutant production, membrane filtration is hindered by surface fouling and the clogging of pores with plastic particles [421]. Therefore, these methods suffer from sustainability issues and have found limited practical usage due to their relatively low efficiency, high operational costs, and the risk of secondary pollution. Addressing the challenge of nano/microplastics in polluted aquatic ecosystems remains a formidable task.

The nascent of thermoplasmonics offers considerable promise as an innovative approach for the effective removal of nano/microplastics [415, 422, 423]. Plasmonic heating employs various mechanisms for nano/microplastics removal, including photothermal decomposition and degradation, enhanced filtration, pollutant detoxification, aggregation, and sedimentation [422, 424–426]. A novel solar-powered technique concentrates sunlight through a glass ball to create a high-density microbubble interface (Fig. 5.22) [415]. This microbubble acts as a convection-driven "furnace" where plastic particles are drawn in, heated and fused due to density differences, effectively transforming them into larger aggregates. These clusters may then settle more easily or be captured by standard treatment processes [417, 422]. As this method uses free solar energy without additional power needs or secondary pollution, it offers a promising solution for managing plastic pollution in aquatic environments.

Thermoplasmonics presents a promising method to address nano/microplastic pollution [260, 424, 427]. However, several challenges must be overcome to effectively utilize plasmonic heating for plastic removal. Critical considerations include ensuring that plasmonic nanostructures specifically target plastic particles without harming the environment, facilitating easy recovery and reuse of the nanoparticles to minimize secondary pollution and costs, and integrating these techniques smoothly into existing water treatment systems for widespread implementation.

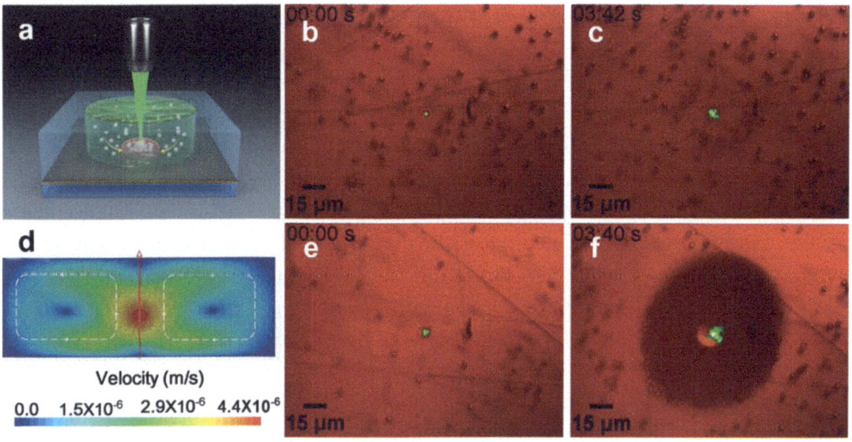


**Fig. 5.22** **a** Collection and fusion of plastic particles in a microbubble induced by a focused laser beam; snapshots showing particle movement in water before **b** and after **c** irradiation without a microbubble. **d** Numerical simulation of convection in water; snapshots depicting plastic particle movement before **e** and after **f** irradiation with microbubble. Reproduced with permission from Ref. [415]. Copyright 2022 Elsevier

## 5.6 Material Science

### *5.6.1 Plasmonic Colloidal Synthesis*

Accurate control of operating temperature, pressure, and reactant concentration is crucial for synthesizing materials with specific functionalities [428, 429]. In particle synthesis, the reaction temperature significantly influences structural attributes such as crystallinity, size, shape, porosity, and surface area [430–432]. Precise control over particle morphology is essential to engineer their optical, electronic, and catalytic properties. By adjusting the reaction temperature, nanoparticles with varied morphologies like rods, cubes, and spheres are synthesized, each exhibiting distinct catalytic activities [433]. Besides conventional heating, temperature gradients can manipulate self-assembly processes, such as forming nematic liquid crystalline structures [434, 435]. Additionally, confining synthesis within spatially restricted areas, termed nanoreactors [436], enables morphology control. Plasmonic heating in metal nanostructures facilitates the creation of these nanoreactors by producing local temperature gradients, essential for nucleation and growth of nanoparticles.

Photothermal heating has been utilized to synthesize various nanomaterials, including carbon nanotubes [437], germanium nanowires [438], and gold nanoellipsoids [439], with specific morphologies resulting from electric and magnetic dipole resonances and hot charge-carrier effects. A light-assisted colloidal synthesis approach using silica-encapsulated gold bipyramids as plasmonic heat sources was also reported for creating iron oxide, silver, and palladium nanoparticles. This method highlighted the critical roles of silica surface chemistry and localized thermal hot spots in promoting nucleation and growth at temperatures lower than those required for conventional heating [440]. The advantages of this approach include the reduced nucleation energy, faster growth kinetics at lower temperatures, and access to diverse superstructures unattainable by traditional methods. However, using nanoscale temperature gradients in thermoplasmonics to synthesize isotropic shells with adjustable chemical compositions and thicknesses down to several nanometers remains challenging. A novel technique leveraging localized plasmonic heating has been developed to synthesize and monitor the in-situ growth of core–shell nanostructures [441]. By adjusting the illumination geometry and chemical environment, this method enables the conformal growth of semiconducting shells such as $CeO_2$, ZnO, and ZnS around plasmonic nanoparticles of varying shapes (Fig. 5.23). The growth rate of the shells correlates with nanoparticle temperature, and the process is monitored in real-time via inelastic light scattering.

It has been assumed that energetic electrons from surface plasmon resonance play a crucial role in photochemical synthesis, though the function of plasmon-generated hot holes in influencing nanostructure morphology remains underexplored. For instance, hot holes cooperatively interact with surface adsorbates to guide the anisotropic expansion of gold nanostructures [442]. These hot holes, stabilized by iodide on surfaces, facilitate selective oxidative etching of $Au^0$, promoting nonuniform growth in different directions to form six-pointed Au nanostars. Moreover, plasmonic nanocrystals demonstrate significantly large asymmetry factors in optical circular dichroism, with hot electron generation displaying pronounced asymmetry,

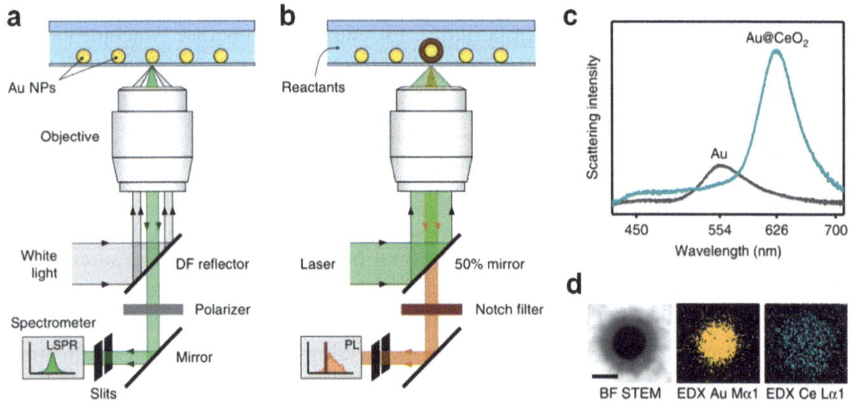


**Fig. 5.23** Photothermal growth of core–shell nanostructures is analyzed using an inverted dark-field microscope. Configuration **a** probes the scattering spectra of individual plasmonic particles under white light, while configuration **b** employs 532 nm continuous-wave laser irradiation to measure the photoluminescence spectra of the emerging core–shell nanoparticle. **c** shows dark-field scattering spectra comparing a single Au particle (gray) to the developed core–shell nanoparticle (teal). **d** presents a BF-STEM image alongside corresponding EDX maps of the particle, with a scale bar of 50 nm. Reproduced from Ref. [441]. Copyright 2020 Nature Publishing Group

thus fostering chiral growth. Leveraging this, chiral plasmonic nanocrystals promote chiral surface growth through energetic electron generation [443]. The asymmetry factors connected with these chiral effects, driven by hot electrons, surpass those seen in chiral molecular photophysics by at least ten times.

Light-driven synthesis of nanostructures, particularly through the localized surface plasmon resonance effect, has emerged as a potent technique in colloidal synthesis [444]. This method leverages localized temperature variations to control reaction kinetics and thermodynamics, facilitating the precise fabrication of nanostructures with customized properties [445–447]. Localized heating accelerates reaction kinetics by applying thermal energy directly at the reaction site, while precise calibration of plasmonic heating creates unique chemical reaction conditions around the nanoparticles. Adjusting light intensity and exposure duration further modulates heat distribution, enabling directional growth and the formation of anisotropic structures like nanorods, nanostars, or nanoplates [435]. However, the role of hot carriers and localized heating in morphological tuning during colloidal synthesis still requires further investigation.

### 5.6.2 Thermoplasmonic Welding

Establishing direct contact between individual nanoobjects is critical for creating hierarchical and multifunctional nanostructures [448, 449]. Nanowelding serves as an effective technique for producing these structures and can be categorized into solid-state bonding, brazing, and fusion welding based on joint physical states [450–453]. Solid-state bonding uses atomic diffusion [454–456] and requires impurity-free weld surfaces with high surface energy [457, 458]. In brazing, the high surface energy of nanomaterials is employed as filler metal to ensure the integrity and functionality of the joined materials [459, 460]. However, a limitation of this technology is that only one joint can be formed at a time, and precise management of nanoscale components is necessary [461]. Proper heat management is essential in nanowelding to maintain material quality, given the high surface energy and sensitivity to heat due to increased surface-to-volume ratios [454]. Fusion methods like Joule heating [462–464], electron beam [465–467], and ultrafast laser welding [468–470] are also employed but risk damaging nanoobjects due to high heat inputs. Consequently, developing precise interconnection technologies to address these challenges is highly desirable.

Thermoplasmonic welding leverages competitive light absorption and heat generation to tune the temperature distribution of metal nanostructures. Metal particles act as nanoheat sources, facilitating self-limited bonding that achieves high precision, low energy use, and scalability [471]. The interaction of light with nanogaps enhances the electromagnetic field, focusing heat precisely where materials need to be joined. Geometry-dependent heating precisely limits the extent of the weld, avoiding damage to substrates with low thermal tolerance [472]. Using pulsed lasers, this technique finely adjusts the temperature of nanostructures with excellent temporal and spatial control [23]. For instance, metallic nanowires can be efficiently assembled into vast

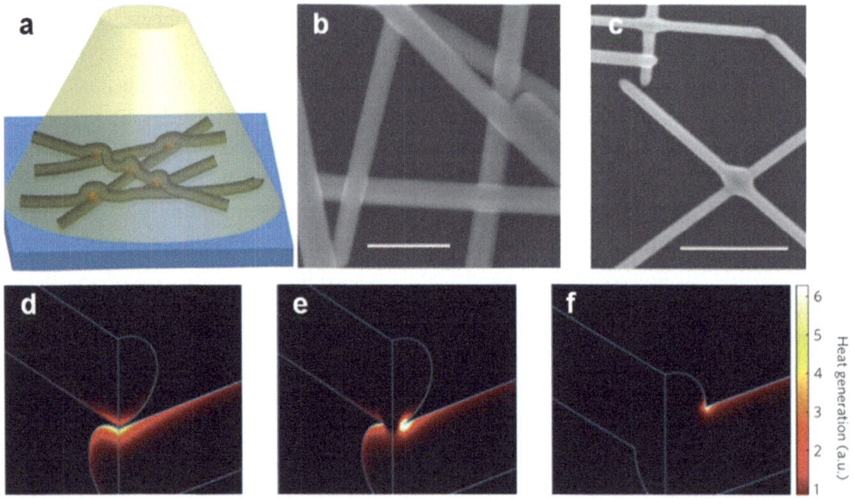


**Fig. 5.24** **a** Thermoplasmonic welding leverages small gaps for intense local heating via strong field concentration (red). **b** SEM image prior to illumination, scale bar: 200 nm. **c** SEM image post-optical welding, scale bar: 500 nm. **d**–**f** Finite element method simulations depict optical heat generation. Reproduced with permission from Ref. [473]. Copyright 2012 Nature Publishing Group

interconnected networks using this method, where naturally occurring small gaps at junctions allow precise heat concentration for welding (Fig. 5.24) [473]. Thermoplasmonic welding offers a noncontact, highly efficient approach, suited to flexible manufacturing systems [474–476]. It has been used to join nanoparticles, nanowires, and ultrathin two-dimensional materials [476, 477], resulting in photoelectrically active NWs used in electronic devices to replace indium tin oxide in transparent electrodes [478]. This welding technology is also integrated with large-scale production techniques like roll-to-roll and inkjet printing [479, 480], indicating its versatility and extensive application potential.

Thermoplasmonic welding, akin to fusion welding, faces several limitations. The technique requires plasmonic heating to originate from the welding materials themselves, with welding quality influenced by factors such as illumination duration, intensity, and interface geometry. Future improvements could address these challenges through the adoption of heterogeneous welding, the development of new structured materials, and the exploration of additional applications. Currently, the focus is primarily on homogeneous junctions. However, interactions between plasmonic materials and other materials like semiconductors and ceramics need further exploration. These hybrid structures composed of heterogeneous materials generate hot spots with variable local field enhancements that can either help or hinder self-limiting welding [481, 482]. A notable advancement has been the use of semiconductors as spacer materials in multi-interface metal–semiconductor structures achieved through thermoplasmonic welding [483]. This method enhances hot spot volume across the entire spacer due to strong plasmon coupling between the metal nanoparticle core and the decorative metal nanoparticle on the semiconductor shell [484, 485].

Plasmonic welding for bottom-up manufacturing is emerging. Future directions may include: expanding the range of materials suitable for nanowelding based on crystal type, lattice compatibility, size, and chemical reactivity; refining assembly methods' technological pathway and process parameters for precise synthesis, connection, and construction of complex nanostructures, along with improved characterization, process feedback, and assembly consistency; and developing strategies for mending nanodevices to ensure their desired functionality, durability, and longevity.

### *5.6.3 Additive Manufacturing*

Thermoplasmonics is enhancing 3D printing or additive manufacturing by diversifying the material options by the integration of nanomaterials for cutting-edge applications [2, 486, 487]. Traditional powder sintering processes, which employ expensive high-power lasers to heat polymers [488] are being supplemented by innovative techniques such as the use of thermoplasmonics. This method uses metal nanoparticles as an "invisible ink" that resonates within the NIR spectrum [489], allowing the creation of colorful prints on this base by adding various dyes [490, 491]. Moreover, plasmonic heating provides refined control over temperature application, essential for working with temperature-sensitive materials to avoid deformation or damage. It also enables the production of intricate geometries by selectively heating specific areas, thus accelerating deposition and solidification processes while enhancing the overall efficiency of the manufacturing process. This approach minimizes the need for extensive post-processing by enabling quicker solidification of the deposited material, thereby reducing production times [2, 282]. The integration of plasmonic particles into printing materials can be tailored to adjust thermal conductivity, mechanical strength, and other properties, offering new prospects for fabricating functional, high-performance components with enhanced features.

Various metallic nanostructures, composed of noble metals like silver, gold, and copper nanoparticles, are encapsulated with organic solvents, serving as stabilizing agents in suitable liquid media for 3D printing applications [282, 489]. For instance, Powell et al. crafted a synthesis process involving gold nanorods coated with silica which were then combined with polyamide powders to create both stable and colorful nanocomposites. Upon illumination, these nanocomposites undergo a resonance response, generating heat conversion. When mixed with dyes, this phenomenon can be harnessed to print colored 3D objects, facilitating the production of vibrant prints (Fig. 5.25) [491].

**Fig. 5.25** Thermoplasmonics in additive manufacturing **a** Utilizing gold nanorods for the sintering of adjacent polymer beads. **b** Production of a colorful sintered object employing PA12 powder combined with an invisible ink comprised of gold nanorods and standard dyes. Adapted with permission from Ref. [491]. Copyright 2020 American Chemical Society

Color printing, an essential element of visual communication for centuries, has been advanced by the introduction of plasmonic color printing [492, 493]. This technique leverages the resonant behavior of electrons on metal particle surfaces. Interaction with light induces collective oscillations of surface plasmons in these nanostructures, producing vivid colors through absorption, scattering, and interference effects [494–496]. The specific colors produced depend on variables such as particle size, shape, material, and arrangement, which determine their resonant frequencies [494]. Metals like gold, silver, and aluminum are utilized due to their favorable plasmonic properties [497]. Recent advancements in fabrication technologies, including electron beam lithography, nanoimprint lithography, and self-assembly, have enabled the creation of nanostructures at the sub-wavelength scale [498, 499]. These developments enhanced the resolution of plasmonic color printing to the theoretical diffraction limit, permitting the production of intricately detailed color images. Additionally, by manipulating these nanostructures, it is possible to access a wide range of the visible spectrum, thus significantly expanding the color gamut beyond the capabilities of traditional CMYK printing systems.

Albeit promising, the incorporation of plasmonics in additive manufacturing requires several considerations. The inherent nonzero absorption within the visible spectrum of plasmonic metals (e.g., attributable to interband transitions or transverse modes in nanorods) restricts the achievable whiteness of prints [490]. Additionally, for applications aimed at the commercial market, concerns regarding the cost-effectiveness, scalability of production, and the nanostructures' long-term stability must be addressed. Ongoing research efforts are dedicated to developing economically viable mass manufacturing techniques and elucidating the degradation mechanisms that affect the durability of plasmonic colors [500]. As this field advances, it is potential to provide sustainable, high-resolution alternatives to conventional color printing technique.

# References


1. A. Andreozzi, M. Iasiello, C. Tucci, An overview of mathematical models and modulated-heating protocols for thermal ablation **52**, 489–541 (2020)
2. G. Baffou, F. Cichos, R. Quidant, Applications and challenges of thermoplasmonics. Nat. Mater. **19**, 946–958 (2020)
3. G. Baffou, R. Quidant, Thermo-plasmonics: using metallic nanostructures as nano-sources of heat. Laser Photonics Rev. **7**, 171–187 (2012)
4. X. Wu, Y. Suo, H. Shi, R. Liu, F. Wu, T. Wang, L. Ma, H. Liu, Z. Cheng, Deep-tissue photothermal therapy using laser illumination at NIR-IIa window. Nano-micro Lett. **12**, 38 (2020)
5. J. Chen, C. Ning, Z. Zhou, P. Yu, Y. Zhu, G. Tan, C. Mao, Nanomaterials as photothermal therapeutic agents. Prog. Mater. Sci. **99**, 1–26 (2019)
6. J. González-Colsa, A. Kuzyk, P. Albella, *On the Photothermal Response of DNA-Au Core/ Shell Nanotoroids as Potential Agents for Photothermal Therapies, Small Structures* (2024)
7. A.B. Bucharskaya, G.N. Maslyakova, M.L. Chekhonatskaya, G.S. Terentyuk, N.A. Navolokin, B.N. Khlebtsov, N.G. Khlebtsov, A.N. Bashkatov, E.A. Genina, V.V. Tuchin, Plasmonic photothermal therapy: approaches to advanced strategy. Lasers Surg. Med. **50**, 1025–1033 (2018)
8. Y. Shen, H.D.A. Santos, E.C. Ximendes, J. Lifante, A. Sanz-Portilla, L. Monge, N. Fernández, I. Chaves-Coira, C. Jacinto, C.D.S. Brites, L.D. Carlos, A. Benayas, M.C. Iglesias-de la Cruz, D. Jaque, $Ag_2S$ nanoheaters with multiparameter sensing for reliable thermal feedback during in vivo tumor therapy. Adv. Funct. Mater. **30** (2020)
9. L.R. Hirsch, R.J. Stafford, J.A. Bankson, S.R. Sershen, B. Rivera, R.E. Price, J.D. Hazle, N.J. Halas, J.L. West, Nanoshell-mediated near-infrared thermal therapy of tumors under magnetic resonance guidance. Proc. Natl. Acad. Sci. U.S.A. **100**, 13549–13554 (2003)
10. H. Chatterjee, D.S. Rahman, M. Sengupta, S.K. Ghosh, Gold nanostars in plasmonic photothermal therapy: the role of tip heads in the thermoplasmonic landscape. J. Phys. Chem. C **122**, 13082–13094 (2018)
11. M. Ha, S.H. Nam, K. Sim, S.E. Chong, J. Kim, Y. Kim, Y. Lee, J.M. Nam, Highly efficient photothermal therapy with cell-penetrating peptide-modified bumpy au triangular nanoprisms using low laser power and low probe dose. Nano Lett. **21**, 731–739 (2021)
12. J. Chen, M. Gong, Y. Fan, J. Feng, L. Han, H.L. Xin, M. Cao, Q. Zhang, D. Zhang, D. Lei, Y. Yin, Collective plasmon coupling in gold nanoparticle clusters for highly efficient photothermal therapy. ACS Nano **16**, 910–920 (2022)
13. X. Li, J.F. Lovell, J. Yoon, X. Chen, Clinical development and potential of photothermal and photodynamic therapies for cancer, nature reviews. Clin. Oncol. **17**, 657–674 (2020)
14. C. Xu, K. Pu, Second near-infrared photothermal materials for combinational nanotheranostics. Chem. Soc. Rev. **50**, 1111–1137 (2021)
15. X. Cui, Q. Ruan, X. Zhuo, X. Xia, J. Hu, R. Fu, Y. Li, J. Wang, H. Xu, Photothermal nanomaterials: a powerful light-to-heat converter. Chem. Rev. **123**, 6891–6952 (2023)
16. C. Ayala-Orozco, C. Urban, M.W. Knight, A.S. Urban, O. Neumann, S.W. Bishnoi, S. Mukherjee, A.M. Goodman, H. Charron, T. Mitchell, M. Shea, R. Roy, S. Nanda, R. Schiff, N.J. Halas, A. Joshi, Au nanomatryoshkas as efficient near-infrared photothermal transducers for cancer treatment: benchmarking against nanoshells. ACS Nano **8**, 6372–6381 (2014)
17. N. Fernandes, C.F. Rodrigues, A.F. Moreira, I.J. Correia, Overview of the application of inorganic nanomaterials in cancer photothermal therapy. Biomaterials science **8**, 2990–3020 (2020)
18. J. Li, W. Zhang, W. Ji, J. Wang, N. Wang, W. Wu, Q. Wu, X. Hou, W. Hu, L. Li, Near infrared photothermal conversion materials: mechanism, preparation, and photothermal cancer therapy applications. J. Mater. Chem. B **9**, 7909–7926 (2021)
19. Y. Ma, Y. Zhang, X. Li, Y. Zhao, M. Li, W. Jiang, X. Tang, J. Dou, L. Lu, F. Wang, Y. Wang, Near-infrared II phototherapy induces deep tissue immunogenic cell death and potentiates cancer immunotherapy. ACS Nano **13**, 11967–11980 (2019)

20. H. Sun, Q. Zhang, J. Li, S. Peng, X. Wang, R. Cai, Near-infrared photoactivated nanomedicines for photothermal synergistic cancer therapy. Nano Today **37**, 101073 (2021)
21. D.S. Wagner, N.A. Delk, E.Y. Lukianova-Hleb, J.H. Hafner, M.C. Farach-Carson, D.O. Lapotko, The in vivo performance of plasmonic nanobubbles as cell theranostic agents in zebrafish hosting prostate cancer xenografts. Biomaterials **31**, 7567–7574 (2010)
22. Y. Wang, K.C. Black, H. Luehmann, W. Li, Y. Zhang, X. Cai, D. Wan, S.Y. Liu, M. Li, P. Kim, Z.Y. Li, L.V. Wang, Y. Liu, Y. Xia, Comparison study of gold nanohexapods, nanorods, and nanocages for photothermal cancer treatment. ACS Nano **7**, 2068–2077 (2013)
23. L. Jauffred, A. Samadi, H. Klingberg, P.M. Bendix, L.B. Oddershede, Plasmonic heating of nanostructures. Chem. Rev. **119**, 8087–8130 (2019)
24. A. Akouibaa, R. Masrour, S. Mordane, M. Benhamou, A. El Assyry, A. Derouiche, Optical and thermo-plasmonic properties of spherical and hollow gold nanoparticles injected in cell organelles. J. Drug Deliv. Sci. Technol. **95**, 105559 (2024)
25. M. Izci, C. Maksoudian, B.B. Manshian, S.J. Soenen, The use of alternative strategies for enhanced nanoparticle delivery to solid tumors. Chem. Rev. **121**, 1746–1803 (2021)
26. E.A. Sykes, J. Chen, G. Zheng, W.C. Chan, Investigating the impact of nanoparticle size on active and passive tumor targeting efficiency. ACS Nano **8**, 5696–5706 (2014)
27. J.T. Jorgensen, K. Norregaard, M. Simon Martin, L.B. Oddershede, A. Kjaer, Non-invasive early response monitoring of nanoparticle-assisted photothermal cancer therapy using (18)F-FDG, (18)F-FLT, and (18)F-FET PET/CT imaging. Nanotheranostics **2**, 201–210 (2018)
28. Yang, D., Yang, G., Yang, P., Lv, R., Gai, S., Li, C., He, F., Lin, J., Assembly of Au plasmonic photothermal agent and iron oxide nanoparticles on ultrathin black phosphorus for targeted photothermal and photodynamic cancer therapy. Adv. Funct. Mater. **27** (2017)
29. J. Zhou, Y. Jiang, S. Hou, P.K. Upputuri, D. Wu, J. Li, P. Wang, X. Zhen, M. Pramanik, K. Pu, H. Duan, Compact plasmonic blackbody for cancer theranosis in the near-infrared II window. ACS Nano **12**, 2643–2651 (2018)
30. Q. Chen, L. Xu, C. Liang, C. Wang, R. Peng, Z. Liu, Photothermal therapy with immune-adjuvant nanoparticles together with checkpoint blockade for effective cancer immunotherapy. Nat. Commun. **7**, 13193 (2016)
31. B. Liu, C. Li, Z. Cheng, Z. Hou, S. Huang, J. Lin, Functional nanomaterials for near-infrared-triggered cancer therapy. Biomater. Sci. **4**, 890–909 (2016)
32. S. Son, J. Kim, J. Kim, B. Kim, J. Lee, Y. Kim, M. Li, H. Kang, J.S. Kim, Cancer therapeutics based on diverse energy sources. Chem. Soc. Rev. **51**, 8201–8215 (2022)
33. A.M.G. edited by Ecaterina Andronescu, nanostructures in therapeutic medicine series: nanostructures for drug delivery, Copyright © 2017 Elsevier Inc. All rights reserved, ISBN: 978-0-323-46143-6 (2017)
34. M.J. Mitchell, M.M. Billingsley, R.M. Haley, M.E. Wechsler, N.A. Peppas, R. Langer, Engineering precision nanoparticles for drug delivery. Nat. Rev. Drug Discovery **20**, 101–124 (2021)
35. B. Luo, W. Wang, Y. Zhao, Y. Zhao, Hot-electron dynamics mediated medical diagnosis and therapy. Chem. Rev. **123**, 10808–10833 (2023)
36. E. Guisasola, A. Baeza, L. Asín, J.M. de la Fuente, M. Vallet-Regí, Heating at the nanoscale through drug-delivery devices: fabrication and synergic effects in cancer treatment with nanoparticles. Small Methods **2** (2018)
37. R. Chen, J. Shi, C. Liu, J. Li, S. Cao, In situ self-assembly of gold nanorods with thermal-responsive microgel for multi-synergistic remote drug delivery. Adv. Compos. Hybrid Mater. **5**, 2223–2234 (2021)
38. C. Wang, M. Vazquez-Gonzalez, M. Fadeev, Y.S. Sohn, R. Nechushtai, I. Willner, Thermoplasmonic-triggered release of loads from DNA-modified hydrogel microcapsules functionalized with Au nanoparticles or Au nanorods. Small **16**, e2000880 (2020)
39. Y. Liu, F. Mo, J. Hu, Q. Jiang, X. Wang, Z. Zou, X.Z. Zhang, D.W. Pang, X. Liu, Precision photothermal therapy and photoacoustic imaging by in situ activatable thermoplasmonics. Chem. Sci. **12**, 10097–10105 (2021)

40. R. Huschka, J. Zuloaga, M.W. Knight, L.V. Brown, P. Nordlander, N.J. Halas, Light-induced release of DNA from gold nanoparticles: nanoshells and nanorods. J. Am. Chem. Soc. **133**, 12247–12255 (2011)
41. R.S. Riley, M.N. Dang, M.M. Billingsley, B. Abraham, L. Gundlach, E.S. Day, Evaluating the mechanisms of light-triggered siRNA release from nanoshells for temporal control over gene regulation. Nano Lett. **18**, 3565–3570 (2018)
42. V.T. Ruhoff, M.R. Arastoo, G. Moreno-Pescador, P.M. Bendix, Biological applications of thermoplasmonics. Nano Lett. (2024)
43. W. Ni, J. Wu, H. Fang, Y. Feng, Y. Hu, L. Lin, J. Chen, F. Chen, H. Tian, Photothermal-chemotherapy enhancing tumor immunotherapy by multifunctional metal-organic framework based drug delivery system. Nano Lett. **21**, 7796–7805 (2021)
44. C. Zhan, Y. Huang, G. Lin, S. Huang, F. Zeng, S. Wu, A Gold Nanocage/cluster hybrid structure for whole-body multispectral optoacoustic tomography imaging EGFR inhibitor delivery, and photothermal therapy. Small **15**, e1900309 (2019)
45. X. Han, J. Huang, H. Lin, Z. Wang, P. Li, Y. Chen, 2D ultrathin MXene-based drug-delivery nanoplatform for synergistic photothermal ablation and chemotherapy of cancer. Adv. Healthcare Mater. **7**, e1701394 (2018)
46. E. Aznar, M. Oroval, L. Pascual, J.R. Murguia, R. Martinez-Manez, F. Sancenon, Gated materials for on-command release of guest molecules. Chem. Rev. **116**, 561–718 (2016)
47. A. Hernandez Montoto, R. Montes, A. Samadi, M. Gorbe, J.M. Terres, R. Cao-Milan, E. Aznar, J. Ibanez, R. Masot, M.D. Marcos, M. Orzaez, F. Sancenon, L.B. Oddershede, R. Martinez-Manez, Gold nanostars coated with mesoporous silica are effective and nontoxic photothermal agents capable of gate keeping and laser-induced drug release. ACS Appl. Mater. Interf. **10**, 27644–27656 (2018)
48. Z. Li, K. Xu, L. Qin, D. Zhao, N. Yang, D. Wang, Y. Yang, Hollow nanomaterials in advanced drug delivery systems: from single- to multiple shells. Adv. Mater. **35**, e2203890 (2023)
49. D. Zhao, Y. Wei, J. Xiong, C. Gao, D. Wang, Response and regulation of the microenvironment based on hollow structured drug delivery systems. Adv. Funct. Mater. **33** (2023)
50. M. Sun, Z. Ji, L. He, C. Zhao, L. Ma, X. Xu, E.J. Cornel, Z. Fan, X. Xu, Instant intracellular delivery of miRNA via photothermal effect induced on plasmonic pyramid arrays. Adv. Funct. Mater. **32** (2021)
51. E. Blanco, H. Shen, M. Ferrari, Principles of nanoparticle design for overcoming biological barriers to drug delivery. Nat. Biotechnol. **33**, 941–951 (2015)
52. R. Xiong, D. Hua, J. Van Hoeck, D. Berdecka, L. Leger, S. De Munter, J.C. Fraire, L. Raes, A. Harizaj, F. Sauvage, G. Goetgeluk, M. Pille, J. Aalders, J. Belza, T. Van Acker, E. Bolea-Fernandez, T. Si, F. Vanhaecke, W.H. De Vos, B. Vandekerckhove, J. van Hengel, K. Raemdonck, C. Huang, S.C. De Smedt, K. Braeckmans, Photothermal nanofibres enable safe engineering of therapeutic cells. Nat. Nanotechnol. **16**, 1281–1291 (2021)
53. Y.C. Wu, T.H. Wu, D.L. Clemens, B.Y. Lee, X.M. Wen, M.A. Horwitz, M.A. Teitell, P.Y. Chiou, Massively parallel delivery of large cargo into mammalian cells with light pulses. Nat. Methods **12**, 439 (2015)
54. U. Bunyatova, M. Dogan, E. Tekin, O. Ferhanoglu, Ultra-stable nano-micro bubbles in a biocompatible medium for safe delivery of anti-cancer drugs. Sci. Rep. **14**, 5321 (2024)
55. J. Gao, J.M. Karp, R. Langer, N. Joshi, The future of drug delivery. Chem. Mater. Publ. Am. Chem. Soc. **35**, 359–363 (2023)
56. S.M. Mirvakili, R. Langer, Wireless on-demand drug delivery. Nat. Electron. **4**, 464–477 (2021)
57. C. Alvarez-Lorenzo, A. Concheiro, Smart drug delivery systems: from fundamentals to the clinic. Chem. Commun. **50**, 7743–7765 (2014)
58. Y. Zhang, F. Liu, Y. Zhang, J. Wang, D. D'Andrea, J.B. Walters, S. Li, H.J. Yoon, M. Wu, S. Li, Z. Hu, T. Wang, J. Choi, K. Bailey, E. Dempsey, K. Zhao, A. Lantsova, Y. Bouricha, I. Huang, H. Guo, X. Ni, Y. Wu, G. Lee, F. Jiang, Y. Huang, C.K. Franz, J.A. Rogers, Self-powered, light-controlled, bioresorbable platforms for programmed drug delivery. Proc. Natl. Acad. Sci. U.S.A. **120**, e2217734120 (2023)

59. H.M.L. Robert, J. Savatier, S. Vial, J. Verghese, B. Wattellier, H. Rigneault, S. Monneret, J. Polleux, G. Baffou, Photothermal control of heat-shock protein expression at the single cell level. Small **14**, e1801910 (2018)
60. G. Baffou, J. Polleux, H. Rigneault, S. Monneret, Super-heating and micro-bubble generation around plasmonic nanoparticles under cw Illumination. J. Phys. Chem. C **118**, 4890–4898 (2014)
61. M. Li, T. Lohmuller, J. Feldmann, Optical injection of gold nanoparticles into living cells. Nano Lett. **15**, 770–775 (2015)
62. X. Zhao, Y. Shi, T. Pan, D. Lu, J. Xiong, B. Li, H. Xin, In Situ single-cell surgery and intracellular organelle manipulation via thermoplasmonics combined optical trapping. Nano Lett. **22**, 402–410 (2022)
63. J. Wang, X. Qu, C. Xu, Z. Zhang, G. Qi, Y. Jin, Thermoplasmonic regulation of the mitochondrial metabolic state for promoting directed differentiation of dental pulp stem cells. Anal. Chem. **94**, 9564–9571 (2022)
64. A. Marino, S. Arai, Y. Hou, A. Degl'Innocenti, V. Cappello, B. Mazzolai, Y.T. Chang, V. Mattoli, M. Suzuki, G. Ciofani, Gold nanoshell-mediated remote myotube activation. ACS Nano **11**, 2494–2508 (2017)
65. A. Andolfi, P. Arnaldi, D.D. Lisa, S. Pepe, M. Frega, A. Fassio, A. Lagazzo, S. Martinoia, L. Pastorino, A micropatterned thermoplasmonic substrate for neuromodulation of in vitro neuronal networks. Acta Biomater. **158**, 281–291 (2023)
66. W. Lu, G. Zhang, R. Zhang, L.G. Flores, Q. Huang, J.G. Gelovani, C. Li, Tumor site-specific silencing of NF-kappaB p65 by targeted hollow gold nanosphere-mediated photothermal transfection. Can. Res. **70**, 3177–3188 (2010)
67. N. Hong, Y. Nam, Thermoplasmonic neural chip platform for in situ manipulation of neuronal connections in vitro. Nat. Commun. **11**, 6313 (2020)
68. A. Sutton, T. Shirman, J.V. Timonen, G.T. England, P. Kim, M. Kolle, T. Ferrante, L.D. Zarzar, E. Strong, J. Aizenberg, Photothermally triggered actuation of hybrid materials as a new platform for in vitro cell manipulation. Nat. Commun. **8**, 14700 (2017)
69. G. Moreno-Pescador, M.R. Arastoo, V.T. Ruhoff, S. Chiantia, R. Daniels, P.M. Bendix, Thermoplasmonic vesicle fusion reveals membrane phase segregation of influenza spike proteins. Nano Lett. **23**, 3377–3384 (2023)
70. A. Rorvig-Lund, A. Bahadori, S. Semsey, P.M. Bendix, L.B. Oddershede, Vesicle fusion triggered by optically heated gold nanoparticles. Nano Lett. **15**, 4183–4188 (2015)
71. D. Yeheskely-Hayon, L. Minai, L. Golan, E.J. Dann, D. Yelin, Optically induced cell fusion using bispecific nanoparticles. Small **9**, 3771–3777 (2013)
72. A. Bahadori, L.B. Oddershede, P.M. Bendix, Hot-nanoparticle-mediated fusion of selected cells. Nano Res. **10**, 2034–2045 (2017)
73. G. Bolognesi, M.S. Friddin, A. Salehi-Reyhani, N.E. Barlow, N.J. Brooks, O. Ces, Y. Elani, Sculpting and fusing biomimetic vesicle networks using optical tweezers. Nat. Commun. **9**, 1882 (2018)
74. Y. Wang, Y. Fei, T. Yang, Z. Luo, Y. Xu, B. Su, X. Lin, Nanotechnology for ultrafast nucleic acid amplification. Nano Today **48**, 101749 (2023)
75. H.L. Zhu, H.Q. Zhang, Y. Xu, S. Lassákova, M. Korabecná, P. Neuzil, PCR past, present and future. Biotechniques **69**, 317–325 (2020)
76. F. Postollec, H. Falentin, S. Pavan, J. Combrisson, D. Sohier, Recent advances in quantitative PCR (qPCR) applications in food microbiology. Food Microbiol. **28**, 848–861 (2011)
77. J. Cheong, H. Yu, C.Y. Lee, J.U. Lee, H.J. Choi, J.H. Lee, H. Lee, J. Cheon, Fast detection of SARS-CoV-2 RNA via the integration of plasmonic thermocycling and fluorescence detection in a portable device. Nat. Biomed. Eng. **4**, 1159–1167 (2020)
78. B.H. Kang, K.W. Jang, E.S. Yu, H. Na, Y.J. Lee, W.Y. Ko, N. Bae, D. Rho, K.H. Jeong, Ultrafast plasmonic nucleic acid amplification and real-time quantification for decentralized molecular diagnostics. ACS Nano **17**, 6507–6518 (2023)
79. S.H. Lee, S.M. Park, B.N. Kim, O.S. Kwon, W.Y. Rho, B.H. Jun, Emerging ultrafast nucleic acid amplification technologies for next-generation molecular diagnostics. Biosens. Bioelectron. **141**, 111448 (2019)

80. B.K. Kim, S.A. Lee, M. Park, E.J. Jeon, M.J. Kim, J.M. Kim, H. Kim, S. Jung, S.K. Kim, Ultrafast real-time PCR in photothermal microparticles. ACS Nano **16**, 20533–20544 (2022)
81. J.H. Son, B. Cho, S. Hong, S.H. Lee, O. Hoxha, A.J. Haack, L.P. Lee, Ultrafast photonic PCR. Light Sci. Appl. **4**, e280–e280 (2015)
82. J. Kim, H. Kim, J.H. Park, S. Jon, Gold nanorod-based photo-PCR system for one-step, rapid detection of bacteria. Nanotheranostics **1**, 178–185 (2017)
83. J.H. Lee, Z. Cheglakov, J. Yi, T.M. Cronin, K.J. Gibson, B. Tian, Y. Weizmann, Plasmonic photothermal gold bipyramid nanoreactors for ultrafast real-time bioassays. J. Am. Chem. Soc. **139**, 8054–8057 (2017)
84. P.J.R. Roche, M. Najih, S.S. Lee, L.K. Beitel, M.L. Carnevale, M. Paliouras, A.G. Kirk, M.A. Trifiro, Real time plasmonic qPCR: how fast is ultra-fast? 30 cycles in 54 seconds. Analyst **142**, 1746–1755 (2017)
85. S. Talebian, G.G. Wallace, A. Schroeder, F. Stellacci, J. Conde, Nanotechnology-based disinfectants and sensors for SARS-CoV-2. Nat. Nanotechnol. **15**, 618–621 (2020)
86. C. Weiss, M. Carriere, L. Fusco, I. Capua, J.A. Regla-Nava, M. Pasquali, J.A. Scott, F. Vitale, M.A. Unal, C. Mattevi, D. Bedognetti, A. Merkoci, E. Tasciotti, A. Yilmazer, Y. Gogotsi, F. Stellacci, L.G. Delogu, Toward nanotechnology-enabled approaches against the COVID-19 pandemic. ACS Nano **14**, 6383–6406 (2020)
87. Z. Tang, X. Zhang, Y. Shu, M. Guo, H. Zhang, W. Tao, Insights from nanotechnology in COVID-19 treatment. Nano Today **36**, 101019 (2021)
88. A. Yakoubi, C.E.B. Dhafer, Advanced plasmonic nanoparticle-based techniques for the prevention, detection, and treatment of current COVID-19. Plasmonics **18**, 311–347 (2023)
89. H. Zhong, Z. Zhu, P. You, J. Lin, C.F. Cheung, V.L. Lu, F. Yan, C.Y. Chan, G. Li, Plasmonic and superhydrophobic self-decontaminating N95 respirators. ACS Nano **14**, 8846–8854 (2020)
90. H. Zhong, Z. Zhu, J. Lin, C.F. Cheung, V.L. Lu, F. Yan, C.Y. Chan, G. Li, Reusable and recyclable graphene masks with outstanding superhydrophobic and photothermal performances. ACS Nano **14**, 6213–6221 (2020)
91. S. Kumar, M. Karmacharya, S.R. Joshi, O. Gulenko, J. Park, G.H. Kim, Y.K. Cho, Photoactive antiviral face mask with self-sterilization and reusability. Nano Lett. **21**, 337–343 (2021)
92. A.K. Bhardwaj, A. Shukla, S. Maurya, S.C. Singh, K.N. Uttam, S. Sundaram, M.P. Singh, R. Gopal, Direct sunlight enabled photo-biochemical synthesis of silver nanoparticles and their bactericidal efficacy: photon energy as key for size and distribution control. J. Photochem. Photobiol. B. Biol. **188**, 42–49 (2018)
93. X. Cai, M. Chen, A. Prominski, Y. Lin, N. Ankenbruck, J. Rosenberg, M. Nguyen, J. Shi, A. Tomatsidou, G. Randall, D. Missiakas, J. Fung, E.B. Chang, P. Penaloza-MacMaster, B. Tian, J. Huang, A multifunctional neutralizing antibody-conjugated nanoparticle inhibits and inactivates SARS-CoV-2. Adv Sci (Weinh) **9**, e2103240 (2022)
94. W. Gong, B. Li, F. She, L. Guo, H. Li, N. Lu, Diffusion-driven height readout analytical device based on gold nanobipyramid-doped supramolecular hydrogel for point-of-care bioanalysis. Sens. Actuators, B Chem. **418**, 136330 (2024)
95. A. Campu, I. Muresan, M. Potara, D.R. Lazar, F.L. Lazar, S. Cainap, D.M. Olinic, D. Maniu, S. Astilean, M. Focsan, Portable microfluidic plasmonic chip for fast real-time cardiac troponin I biomarker thermoplasmonic detection. J. Mater. Chem. B. (2023)
96. J. Homola, Surface plasmon resonance sensors for detection of chemical and biological species. Chem. Rev. **108**, 462–493 (2008)
97. K.M. Mayer, J.H. Hafner, Localized surface plasmon resonance sensors. Chem. Rev. **111**, 3828–3857 (2011)
98. Y. Xiao, Z. Zhang, S. Yin, X. Ma, Nanoplasmonic biosensors for precision medicine. Front. Chem. **11**, 1209744 (2023)
99. C. Zhan, X.J. Chen, J. Yi, J.F. Li, D.Y. Wu, Z.Q. Tian, From plasmon-enhanced molecular spectroscopy to plasmon-mediated chemical reactions. Nat. Rev. Chem. **2**, 216–230 (2018)
100. N. Liu, M.L. Tang, M. Hentschel, H. Giessen, A.P. Alivisatos, Nanoantenna-enhanced gas sensing in a single tailored nanofocus. Nat. Mater. **10**, 631–636 (2011)

101. M. Mesch, B. Metzger, M. Hentschel, H. Giessen, Nonlinear plasmonic sensing. Nano Lett. **16**, 3155–3159 (2016)
102. M. Riskin, R. Tel-Vered, O. Lioubashevski, I. Willner, Ultrasensitive surface plasmon resonance detection of trinitrotoluene by a bis-aniline-cross-linked Au nanoparticles composite. J. Am. Chem. Soc. **131**, 7368–7378 (2009)
103. L. Song, J. Chen, B.B. Xu, Y. Huang, Flexible plasmonic biosensors for healthcare monitoring: progress and prospects. ACS Nano **15**, 18822–18847 (2021)
104. G.C. Phan-Quang, H.K. Lee, H.W. Teng, C.S.L. Koh, B.Q. Yim, E.K.M. Tan, W.L. Tok, I.Y. Phang, X.Y. Ling, Plasmonic hotspots in air: an omnidirectional three-dimensional platform for stand-off in-air SERS sensing of airborne species. Angew. Chem. **57**, 5792–5796 (2018)
105. Y.J. Zhang, P.M. Radjenovic, X.S. Zhou, H. Zhang, J.L. Yao, J.F. Li, Plasmonic core-shell nanomaterials and their applications in spectroscopies. Adv. Mater. **33**, e2005900 (2021)
106. G. Qiu, Z. Gai, L. Saleh, J. Tang, T. Gui, G.A. Kullak-Ublick, J. Wang, Thermoplasmonic-assisted cyclic cleavage amplification for self-validating plasmonic detection of SARS-CoV-2. ACS Nano **15**, 7536–7546 (2021)
107. Q. Zhao, J. Piao, W. Peng, J. Wang, W. Gao, X. Wu, H. Wang, X. Gong, J. Chang, B. Zhang, A metal chelator as a plasmonic signal-generation superregulator for ultrasensitive colorimetric bioassays of disease biomarkers. Adv. Sci. **5**, 1800295 (2018)
108. S. Yan, S. Shan, J. Wen, J. Li, N. Kang, Z. Wu, J. Lombardi, H.W. Cheng, J. Wang, J. Luo, N. He, D. Mott, L. Wang, Q. Ge, B.S. Hsiao, M. Poliks, C.J. Zhong, Surface-mediated interconnections of nanoparticles in cellulosic fibrous materials toward 3D sensors. Adv. Mater. **32**, e2002171 (2020)
109. W.T. Koo, Y. Kim, S. Savagatrup, B. Yoon, I. Jeon, S.J. Choi, I.D. Kim, T.M. Swager, Porous ion exchange polymer matrix for ultrasmall Au nanoparticle-decorated carbon nanotube chemiresistors. Chem. Mater. **31**, 5413–5420 (2019)
110. C. Tang, K.H. Ku, S.X. Lennon Luo, A. Concellon, Y.M. Wu, R.Q. Lu, T.M. Swager, Chelating phosphine ligand stabilized AuNPs in methane detection. ACS Nano **14**, 11605–11612 (2020)
111. G. Palermo, G. Strangi, Thermoplasmonic-biosensing demonstration based on the photothermal response of metallic nanoparticles. J. Appl. Phys. **128** (2020)
112. M. Virk, K. Xiong, M. Svedendahl, M. Kall, A.B. Dahlin, A thermal plasmonic sensor platform: resistive heating of nanohole arrays. Nano Lett. **14**, 3544–3549 (2014)
113. Y. Liu, H. Ye, H. Huynh, C. Xie, P. Kang, J.S. Kahn, Z. Qin, Digital plasmonic nanobubble detection for rapid and ultrasensitive virus diagnostics. Nat. Commun. **13**, 1687 (2022)
114. G. Qiu, Z. Gai, Y. Tao, J. Schmitt, G.A. Kullak-Ublick, J. Wang, Dual-functional plasmonic photothermal biosensors for highly accurate severe acute respiratory syndrome coronavirus 2 detection. ACS Nano **14**, 5268–5277 (2020)
115. D. Zopf, A. Pittner, A. Dathe, N. Grosse, A. Csaki, K. Arstila, J.J. Toppari, W. Schott, D. Dontsov, G. Uhlrich, W. Fritzsche, O. Stranik, Plasmonic nanosensor array for multiplexed DNA-based pathogen detection. ACS Sens. **4**, 335–343 (2019)
116. F. Haider, R. Ahmmed Aoni, R. Ahmed, G. Amouzad Mahdiraji, M. Fahmi Azman, F.R.M. Adikan, Mode-multiplex plasmonic sensor for multi-analyte detection. Opt. Lett. **45**, 3945–3948 (2020)
117. A.B. Taylor, P. Zijlstra, Single-molecule plasmon sensing: current status and future prospects. ACS Sens. **2**, 1103–1122 (2017)
118. X. Bi, D.M. Czajkowsky, Z. Shao, J. Ye, Digital colloid-enhanced Raman spectroscopy by single-molecule counting. Nature (2024)
119. C. Lee, B. Lawrie, R. Pooser, K.G. Lee, C. Rockstuhl, M. Tame, Quantum plasmonic sensors. Chem. Rev. **121**, 4743–4804 (2021)
120. J. Qin, S. Jiang, Z. Wang, X. Cheng, B. Li, Y. Shi, D.P. Tsai, A.Q. Liu, W. Huang, W. Zhu, Metasurface micro/nano-optical sensors: principles and applications. ACS Nano **16**, 11598–11618 (2022)
121. D. Boyer, P. Tamarat, A. Maali, B. Lounis, M. Orrit, Photothermal imaging of nanometer-sized metal particles among scatterers. Science **297**, 1160–1163 (2002)

122. L. Cognet, C. Tardin, D. Boyer, D. Choquet, P. Tamarat, B. Lounis, Single metallic nanoparticle imaging for protein detection in cells. Proc. Natl. Acad. Sci. U.S.A. **100**, 11350–11355 (2003)
123. S. Berciaud, L. Cognet, G.A. Blab, B. Lounis, Photothermal heterodyne imaging of individual nonfluorescent nanoclusters and nanocrystals. Phys. Rev. Lett. **93**, 257402 (2004)
124. S. Adhikari, P. Spaeth, A. Kar, M.D. Baaske, S. Khatua, M. Orrit, Photothermal microscopy: imaging the optical absorption of single nanoparticles and single molecules. ACS Nano **14**, 16414–16445 (2020)
125. P. Vermeulen, L. Cognet, B. Lounis, Photothermal microscopy: optical detection of small absorbers in scattering environments. J. Microsc. **254**, 115–121 (2014)
126. K.A. Willets, A.J. Wilson, V. Sundaresan, P.B. Joshi, Super-Resolution Imaging and Plasmonics. Chem. Rev. **117**, 7538–7582 (2017)
127. S.A. Hosseini Jebeli, C.A. West, S.A. Lee, H.J. Goldwyn, C.R. Bilchak, Z. Fakhraai, K.A. Willets, S. Link, D.J. Masiello, Wavelength-dependent photothermal imaging probes nanoscale temperature differences among subdiffraction coupled plasmonic nanorods. Nano Lett. **21**, 5386–5393 (2021)
128. Y. Zhang, H. Zong, C. Zong, Y. Tan, M. Zhang, Y. Zhan, J.X. Cheng, Fluorescence-detected mid-infrared photothermal microscopy. J. Am. Chem. Soc. **143**, 11490–11499 (2021)
129. L. Cognet, S. Berciaud, D. Lasne, B. Lounis, Photothermal methods for single nonluminescent nano-objects. Anal. Chem. **80**, 2288–2294 (2008)
130. R.W. Taylor, V. Sandoghdar, Interferometric scattering microscopy: seeing single nanoparticles and molecules via Rayleigh scattering. Nano Lett. **19**, 4827–4835 (2019)
131. M. Selmke, A. Heber, M. Braun, F. Cichos, Photothermal single particle microscopy using a single laser beam. Appl. Phys. Lett. **105** (2014)
132. G. Baffou, C. Girard, R. Quidant, Mapping heat origin in plasmonic structures. Phys. Rev. Lett. **104**, 136805 (2010)
133. Z.X. Chen, X.N. Shan, Y. Guan, S.P. Wang, J.J. Zhu, N.J. Tao, Imaging local heating and thermal diffusion of nanomaterials with plasmonic thermal microscopy. ACS Nano **9**, 11574–11581 (2015)
134. M. Yorulmaz, S. Nizzero, A. Hoggard, L.Y. Wang, Y.Y. Cai, M.N. Su, W.S. Chang, S. Link, Single-particle absorption spectroscopy by photothermal contrast. Nano Lett. **15**, 3041–3047 (2015)
135. M. Kim, J.H. Lee, J.M. Nam, Plasmonic photothermal nanoparticles for biomedical applications. Adv Sci (Weinh) **6**, 1900471 (2019)
136. W. Li, X. Chen, Gold nanoparticles for photoacoustic imaging. Nanomedicine **10**, 299–320 (2015)
137. I.C. Sun, D.S. Dumani, S.Y. Emelianov, Applications of the photocatalytic and photoacoustic properties of gold nanorods in contrast-enhanced ultrasound and photoacoustic imaging. ACS Nano **18**, 3575–3582 (2024)
138. R. García-Álvarez, L. Chen, A. Nedilko, A. Sánchez-Iglesias, A. Rix, W. Lederle, V. Pathak, T. Lammers, G. von Plessen, K. Kostarelos, L.M. Liz-Marzán, A.J.C. Kuehne, D.N. Chigrin, Optimizing the geometry of photoacoustically active gold nanoparticles for biomedical imaging. ACS Photon. **7**, 646–652 (2020)
139. C. Tian, W. Qian, X. Shao, Z. Xie, X. Cheng, S. Liu, Q. Cheng, B. Liu, X. Wang, Plasmonic nanoparticles with quantitatively controlled bioconjugation for photoacoustic imaging of live cancer cells. Adv. Sci. **3** (2016)
140. T. Repenko, A. Rix, A. Nedilko, J. Rose, A. Hermann, R. Vinokur, S. Moli, R. Cao-Milàn, M. Mayer, G. von Plessen, A. Fery, L. De Laporte, W. Lederle, D.N. Chigrin, A.J.C. Kuehne, Strong photoacoustic signal enhancement by coating gold nanoparticles with melanin for biomedical imaging. Adv. Funct. Mater. **28** (2017)
141. Y. Mantri, J.V. Jokerst, Engineering plasmonic nanoparticles for enhanced photoacoustic imaging. ACS Nano **14**, 9408–9422 (2020)
142. C. Zhou, L. Zhang, T. Sun, Y. Zhang, Y. Liu, M. Gong, Z. Xu, M. Du, Y. Liu, G. Liu, D. Zhang, Activatable NIR-II plasmonic nanotheranostics for efficient photoacoustic imaging and photothermal cancer therapy. Adv. Mater. **33** (2020)

143. Y. Liu, J. He, K. Yang, C. Yi, Y. Liu, L. Nie, N.M. Khashab, X. Chen, Z. Nie, Folding up of gold nanoparticle strings into plasmonic vesicles for enhanced photoacoustic imaging. Angew. Chem. **54**, 15809–15812 (2015)
144. S. Kumar, J. Aaron, K. Sokolov, Directional conjugation of antibodies to nanoparticles for synthesis of multiplexed optical contrast agents with both delivery and targeting moieties. Nat. Protoc. **3**, 314–320 (2008)
145. B. Wang, E. Yantsen, T. Larson, A.B. Karpiouk, S. Sethuraman, J.L. Su, K. Sokolov, S.Y. Emelianov, Plasmonic intravascular photoacoustic imaging for detection of macrophages in atherosclerotic plaques. Nano Lett. **9**, 2212–2217 (2009)
146. K. Wilson, K. Homan, S. Emelianov, Biomedical photoacoustics beyond thermal expansion using triggered nanodroplet vaporization for contrast-enhanced imaging, Nat. Commun. **3** (2012)
147. A.J. Dixon, S. Hu, A.L. Klibanov, J.A. Hossack, Oscillatory dynamics and in vivo photoacoustic imaging performance of plasmonic nanoparticle-coated microbubbles. Small **11**, 3066–3077 (2015)
148. D. Lee, E.J. Park, S.E. Lee, S.H. Jeong, J.Y. Lee, E. Kang, Energy-absorbing and local plasmonic nanodiamond/gold nanocomposites for sustained and enhanced photoacoustic imaging. ACS Sustain. Chem. Eng. **5**, 8284–8293 (2017)
149. W.J. Lee, E.Y. Park, D. Choi, D. Lee, J. Koo, J.G. Min, Y. Jung, S.B. Hong, K. Kim, C. Kim, S. Kim, Colloidal porous AuAg alloyed nanoparticles for enhanced photoacoustic imaging. ACS Appl. Mater. Interfaces **12**, 32270–32277 (2020)
150. L. Li, S. Li, Z. Fan, G. Huang, J. Tang, L. Nie, Current strategies of photoacoustic imaging assisted cancer theragnostics toward clinical studies. ACS Photon. **9**, 2555–2578 (2022)
151. R.M. Abraham-Ekeroth, F. De Angelis, Radioplasmonics: plasmonic transducers in the radiofrequency regime for resonant thermo-acoustic imaging in deep tissues. ACS Photon. **8**, 238–246 (2021)
152. M.T. Kief, R.H. Victora, Materials for heat-assisted magnetic recording. MRS Bull. **43**, 87–92 (2018)
153. A. Nordrum, The fight for the future of the disk drive. IEEE Spectr. **56**, 44–47 (2019)
154. N. Zhou, X. Xu, A.T. Hammack, B.C. Stipe, K. Gao, W. Scholz, E.C. Gage, Plasmonic near-field transducer for heat-assisted magnetic recording. Nanophotonics **3**, 141–155 (2014)
155. R. Katayama, J. Chen, S. Sugiura, Simulation on double-ring-resonator-type device considering gain for heat-assisted magnetic recording. Opt. Rev. **30**, 361–375 (2023)
156. W.A. Challener, C. Peng, A.V. Itagi, D. Karns, W. Peng, Y. Peng, X. Yang, X. Zhu, N.J. Gokemeijer, Y.T. Hsia, G. Ju, R.E. Rottmayer, M.A. Seigler, E.C. Gage, Heat-assisted magnetic recording by a near-field transducer with efficient optical energy transfer. Nat. Photonics **3**, 220–224 (2009)
157. B.V. Budaev, D.B. Bogy, On the lifetime of plasmonic transducers in heat assisted magnetic recording. J. Appl. Phys. **112** (2012)
158. K. Nakagawa, A. Tajiri, K. Tamura, S. Toriumi, Y. Ashizawa, A. Tsukamoto, A. Itoh, Y. Sasaki, S. Saito, M. Takahashi, S. Ohnuki, Thermally assisted magnetic recording applying optical near field with ultra short-time heating. J. Magnet. Soc. Jpn 119–122 (2013)
159. B.C. Stipe, T.C. Strand, C.C. Poon, H. Balamane, T.D. Boone, J.A. Katine, J.-L. Li, V. Rawat, H. Nemoto, A. Hirotsune, O. Hellwig, R. Ruiz, E. Dobisz, D.S. Kercher, N. Robertson, T.R. Albrecht, B.D. Terris, Magnetic recording at 1.5 Pb m−2 using an integrated plasmonic antenna. Nat. Photon. **4**, 484–488 (2010)
160. T. Matsumoto, F. Akagi, M. Mochizuki, H. Miyamoto, B. Stipe, Integrated head design using a nanobeak antenna for thermally assisted magnetic recording. Opt. Express **20**, 18946–18954 (2012)
161. H. Chen, K. Liu, L. Hu, A.A. Al-Ghamdi, X. Fang, New concept ultraviolet photodetectors. Mater. Today **18**, 493–502 (2015)
162. M. Garin, J. Heinonen, L. Werner, T.P. Pasanen, V. Vahanissi, A. Haarahiltunen, M.A. Juntunen, H. Savin, Black-silicon ultraviolet photodiodes achieve external quantum efficiency above 130. Phys. Rev. Lett. **125**, 117702 (2020)

163. M.A. Juntunen, J. Heinonen, V. Vähänissi, P. Repo, D. Valluru, H. Savin, Near-unity quantum efficiency of broadband black silicon photodiodes with an induced junction. Nat. Photonics **10**, 777–781 (2016)
164. C. Zhang, Y. Luo, S.A. Maier, X. Li, Recent progress and future opportunities for hot carrier photodetectors: from ultraviolet to infrared bands. Laser Photon. Rev. **16** (2022)
165. K.T. Lin, H. Lin, B. Jia, Plasmonic nanostructures in photodetection, energy conversion and beyond. Nanophotonics **9**, 3135–3163 (2020)
166. Y.S. Zhu, H.X. Xu, P. Yu, Z.M. Wang, Engineering plasmonic hot carrier dynamics toward efficient photodetection. Appl. Phys. Rev. **8** (2021)
167. A. Dorodnyy, Y. Salamin, P. Ma, J. Vukajlovic Plestina, N. Lassaline, D. Mikulik, P. Romero-Gomez, A. Fontcuberta i Morral, J. Leuthold, Plasmonic photodetectors. IEEE J. Select. Top. Quantum Electron. **24**, 1–13 (2018)
168. M.W. Knight, H. Sobhani, P. Nordlander, N.J. Halas, Photodetection with active optical antennas. Science **332**, 702–704 (2011)
169. C. Scales, P. Berini, Thin-film schottky barrier photodetector models. Ieee J Quantum Elect **46**, 633–643 (2010)
170. C. Zhang, K. Wu, V. Giannini, X. Li, Planar hot-electron photodetection with tamm plasmons. ACS Nano **11**, 1719–1727 (2017)
171. A. Safaei, S. Chandra, M.W. Shabbir, M.N. Leuenberger, D. Chanda, Dirac plasmon-assisted asymmetric hot carrier generation for room-temperature infrared detection. Nat. Commun. **10** (2019)
172. M. Abbasi, C.I. Evans, L. Chen, D. Natelson, Single metal photodetectors using plasmonically-active asymmetric gold nanostructures. ACS Nano **14**, 17535–17542 (2020)
173. S. Kunwar, S. Pandit, J.H. Jeong, J. Lee, Improved photoresponse of UV photodetectors by the incorporation of plasmonic nanoparticles on GaN through the resonant coupling of localized surface plasmon resonance. Nano-micro Lett. **12** (2020)
174. W. Li, J. Valentine, Metamaterial perfect absorber based hot electron photodetection. Nano Lett. **14**, 3510–3514 (2014)
175. Z. Wang, X. Wang, J. Liu, An efficient nanophotonic hot electron solar-blind UV detector. ACS Photon. **5**, 3989–3995 (2018)
176. H. Chalabi, D. Schoen, M.L. Brongersma, Hot-electron photodetection with a plasmonic nanostripe antenna. Nano Lett. **14**, 1374–1380 (2014)
177. D.A. Bandurin, D. Svintsov, I. Gayduchenko, S.G. Xu, A. Principi, M. Moskotin, I. Tretyakov, D. Yagodkin, S. Zhukov, T. Taniguchi, K. Watanabe, I.V. Grigorieva, M. Polini, G.N. Goltsman, A.K. Geim, G. Fedorov, Resonant terahertz detection using graphene plasmons. Nat. Commun. **9**, 5392 (2018)
178. W. Li, J.G. Valentine, Harvesting the loss: surface plasmon-based hot electron photodetection. Nanophotonics **6**, 177–191 (2017)
179. F. Yi, H. Zhu, J.C. Reed, E. Cubukcu, Plasmonically enhanced thermomechanical detection of infrared radiation. Nano Lett. **13**, 1638–1643 (2013)
180. F. Yi, H. Zhu, J.C. Reed, A.Y. Zhu, E. Cubukcu, Thermoplasmonic membrane-based infrared detector. IEEE Photonic Tech. L. **26**, 202–205 (2014)
181. A.A.A. Muheeb Ahmad Alkhalayfeh, Mohd Zamir Pakhuruddin, An overview of enhanced polymer solar cells with embedded plasmonic nanoparticles. Renew. Sustain. Energy Rev. **141**, 110726 (2021)
182. D. Shin, G. Kang, P. Gupta, S. Behera, H. Lee, A.M. Urbas, W. Park, K. Kim, Thermoplasmonic and photothermal metamaterials for solar energy applications. Adv. Opt. Mater. **6**, 1800317 (2018)
183. C. Kuppe, K.R. Rusimova, L. Ohnoutek, D. Slavov, V.K. Valev, "Hot" in plasmonics: temperature-related concepts and applications of metal nanostructures. Adv. Opt. Mater. **8**, 1901166 (2019)
184. S.K. Cushing, N. Wu, Progress and perspectives of plasmon-enhanced solar energy conversion. J. Phys. Chem. Lett. **7**, 666–675 (2016)

185. W.R. Erwin, H.F. Zarick, E.M. Talbert, R. Bardhan, Light trapping in mesoporous solar cells with plasmonic nanostructures. Energy Environ. Sci. **9**, 1577–1601 (2016)
186. H.A. Atwater, A. Polman, Plasmonics for improved photovoltaic devices. Nat. Mater. **9**, 205–213 (2010)
187. Y.H. Jang, Y.J. Jang, S. Kim, L.N. Quan, K. Chung, D.H. Kim, Plasmonic solar cells: from rational design to mechanism overview. Chem. Rev. **116**, 14982–15034 (2016)
188. F. Enrichi, A. Quandt, G.C. Righini, Plasmonic enhanced solar cells: summary of possible strategies and recent results. Renew. Sustain. Energy Rev. **82**, 2433–2439 (2018)
189. R.K. Kashyap, P.P. Pillai, Plasmonic nanoparticles boost solar-to-electricity generation at ambient conditions. Nano Lett. **24**, 5585–5592 (2024)
190. K. Ueno, T. Oshikiri, Q. Sun, X. Shi, H. Misawa, Solid-state plasmonic solar cells. Chem. Rev. **118**, 2955–2993 (2018)
191. Y. Li, W. Li, T. Han, X. Zheng, J. Li, B. Li, S. Fan, C.W. Qiu, Transforming heat transfer with thermal metamaterials and devices. Nat. Rev. Mater. **6**, 488–507 (2021)
192. C.C. Chang, W.J.M. Kort-Kamp, J. Nogan, T.S. Luk, A.K. Azad, A.J. Taylor, D.A.R. Dalvit, M. Sykora, H.T. Chen, High-temperature refractory metasurfaces for solar thermophotovoltaic energy harvesting. Nano Lett. **18**, 7665–7673 (2018)
193. D. Fan, T. Burger, S. McSherry, B. Lee, A. Lenert, S.R. Forrest, Near-perfect photon utilization in an air-bridge thermophotovoltaic cell. Nature **586**, 237–241 (2020)
194. A. Lenert, D.M. Bierman, Y. Nam, W.R. Chan, I. Celanovic, M. Soljacic, E.N. Wang, A nanophotonic solar thermophotovoltaic device. Nat. Nanotechnol. **9**, 126–130 (2014)
195. B. Yang, C. Li, Z. Wang, Q. Dai, Thermoplasmonics in solar energy conversion: materials nanostructured designs, and applications. Adv. Mater. **34**, e2107351 (2022)
196. A. LaPotin, K.L. Schulte, M.A. Steiner, K. Buznitsky, C.C. Kelsall, D.J. Friedman, E.J. Tervo, R.M. France, M.R. Young, A. Rohskopf, S. Verma, E.N. Wang, A. Henry, Thermophotovoltaic efficiency of 40. Nature **604**, 287–291 (2022)
197. T. Krekeler, S.S. Rout, G.V. Krishnamurthy, M. Störmer, M. Arya, A. Ganguly, D.S. Sutherland, S.I. Bozhevolnyi, M. Ritter, K. Pedersen, A.Y. Petrov, M. Eich, M. Chirumamilla, Unprecedented thermal stability of plasmonic titanium nitride films up to 1400 °C. Adv. Opt. Mater. (2021) 2100323
198. V. Rinnerbauer, A. Lenert, D.M. Bierman, Y.X. Yeng, W.R. Chan, R.D. Geil, J.J. Senkevich, J.D. Joannopoulos, E.N. Wang, M. Soljačić, I. Celanovic, Metallic photonic crystal absorber-emitter for efficient spectral control in high-temperature solar thermophotovoltaics. Adv. Energy Mater. **4**, 1400334 (2014)
199. A. Fiorino, L. Zhu, D. Thompson, R. Mittapally, P. Reddy, E. Meyhofer, Nanogap near-field thermophotovoltaics. Nat. Nanotechnol. **13**, 806–811 (2018)
200. G.T. Papadakis, S. Buddhiraju, Z. Zhao, B. Zhao, S. Fan, Broadening near-field emission for performance enhancement in thermophotovoltaics. Nano Lett. **20**, 1654–1661 (2020)
201. R. St-Gelais, G.R. Bhatt, L. Zhu, S. Fan, M. Lipson, Hot carrier-based near-field thermophotovoltaic energy conversion. ACS Nano **11**, 3001–3009 (2017)
202. X. Liu, R.Z. Zhang, Z. Zhang, Near-perfect photon tunneling by hybridizing graphene plasmons and hyperbolic modes. ACS Photon. **1**, 785–789 (2014)
203. M.R.S. Dias, T. Gong, M.A. Duncan, S.C. Ness, S.J. McCormack, M.S. Leite, J.N. Munday, Photonics roadmap for ultra-high-temperature thermophotovoltaics. Joule **7**, 2209–2227 (2023)
204. A. Lenert, S.R. Forrest, Nexus of solar and thermal photovoltaic technology could help solve the energy storage problem. Joule **6**, 1144–1147 (2022)
205. M. Gao, L. Zhu, C.K. Peh, G.W. Ho, Solar absorber material and system designs for photothermal water vaporization towards clean water and energy production. Energy Environ. Sci. **12**, 841–864 (2019)
206. G. Liu, T. Chen, J. Xu, G. Yao, J. Xie, Y. Cheng, Z. Miao, K. Wang, Salt-rejecting solar interfacial evaporation. Cell Rep. Phys. Sci. **2**, 100310 (2021)
207. Z. Wang, T. Horseman, A.P. Straub, N.Y. Yip, D. Li, M. Elimelech, S. Lin, Pathways and challenges for efficient solar-thermal desalination. Sci. Adv. **5** (2019) eaax0763

208. L. Zhang, Z. Xu, L. Zhao, B. Bhatia, Y. Zhong, S. Gong, E.N. Wang, Passive, high-efficiency thermally-localized solar desalination. Energy Environ. Sci. **14**, 1771–1793 (2021)
209. P. Tao, G. Ni, C. Song, W. Shang, J. Wu, J. Zhu, G. Chen, T. Deng, Solar-driven interfacial evaporation. Nat. Energy **3**, 1031–1041 (2018)
210. Z. Xu, Z. Li, Y. Jiang, G. Xu, M. Zhu, W.C. Law, K.T. Yong, Y. Wang, C. Yang, B. Dong, F. Xing, Recent advances in solar-driven evaporation systems. J. Mater. Chem. A **8**, 25571–25600 (2020)
211. G. Liu, J. Xu, K. Wang, Solar water evaporation by black photothermal sheets. Nano Energy **41**, 269–284 (2017)
212. A. Politano, P. Argurio, G. Di Profio, V. Sanna, A. Cupolillo, S. Chakraborty, H.A. Arafat, E. Curcio, Photothermal membrane distillation for seawater desalination. Adv. Mater. **29** (2017)
213. O. Neumann, A.S. Urban, J. Day, S. Lal, P. Nordlander, N.J. Halas, Solar vapor generation enabled by nanoparticles. ACS Nano **7**, 42–49 (2013)
214. M.S. Zielinski, J.W. Choi, T. La Grange, M. Modestino, S.M. Hashemi, Y. Pu, S. Birkhold, J.A. Hubbell, D. Psaltis, Hollow mesoporous plasmonic nanoshells for enhanced solar vapor generation. Nano Lett. **16**, 2159–2167 (2016)
215. Z. Fang, Y.R. Zhen, O. Neumann, A. Polman, F.J. Garcia de Abajo, P. Nordlander, N.J. Halas, Evolution of light-induced vapor generation at a liquid-immersed metallic nanoparticle. Nano Lett. **13**, 1736–1742 (2013)
216. M. Gao, P.K.N. Connor, G.W. Ho, Plasmonic photothermic directed broadband sunlight harnessing for seawater catalysis and desalination. Energy Environ. Sci. **9**, 3151–3160 (2016)
217. Z. Wang, Y. Liu, P. Tao, Q. Shen, N. Yi, F. Zhang, Q. Liu, C. Song, D. Zhang, W. Shang, T. Deng, Bio-inspired evaporation through plasmonic film of nanoparticles at the air-water interface. Small **10**, 3234–3239 (2014)
218. X. Wang, Y. He, X. Liu, G. Cheng, J. Zhu, Solar steam generation through bio-inspired interface heating of broadband-absorbing plasmonic membranes. Appl. Energy **195**, 414–425 (2017)
219. L. Zhou, Y. Tan, J. Wang, W. Xu, Y. Yuan, W. Cai, S. Zhu, J. Zhu, 3D self-assembly of aluminium nanoparticles for plasmon-enhanced solar desalination. Nat. Photonics **10**, 393–398 (2016)
220. Y. Liu, S. Yu, R. Feng, A. Bernard, Y. Liu, Y. Zhang, H. Duan, W. Shang, P. Tao, C. Song, T. Deng, A bioinspired, reusable, paper-based system for high-performance large-scale evaporation. Adv. Mater. **27**, 2768–2774 (2015)
221. H. Liu, X. Zhang, Z. Hong, Z. Pu, Q. Yao, J. Shi, G. Yang, B. Mi, B. Yang, X. Liu, H. Jiang, X. Hu, A bioinspired capillary-driven pump for solar vapor generation. Nano Energy **42**, 115–121 (2017)
222. K. Bae, G. Kang, S.K. Cho, W. Park, K. Kim, W.J. Padilla, Flexible thin-film black gold membranes with ultrabroadband plasmonic nanofocusing for efficient solar vapour generation. Nat. Commun. **6**, 10103 (2015)
223. L. Zhou, Y. Tan, D. Ji, B. Zhu, P. Zhang, J. Xu, Q. Gan, Z. Yu, J. Zhu, Self-assembly of highly efficient, broadband plasmonic absorbers for solar steam generation. Sci. Adv. **2**, e1501227 (2016)
224. L. Mascaretti, A. Schirato, R. Zbořil, Š Kment, P. Schmuki, A. Alabastri, A. Naldoni, Solar steam generation on scalable ultrathin thermoplasmonic TiN nanocavity arrays. Nano Energy **83**, 105828 (2021)
225. J. Chen, J. Feng, Z. Li, P. Xu, X. Wang, W. Yin, M. Wang, X. Ge, Y. Yin, Space-confined seeded growth of black silver nanostructures for solar steam generation. Nano Lett. **19**, 400–407 (2019)
226. M. Zhu, Y. Li, F. Chen, X. Zhu, J. Dai, Y. Li, Z. Yang, X. Yan, J. Song, Y. Wang, E. Hitz, W. Luo, M. Lu, B. Yang, L. Hu, Plasmonic wood for high-efficiency solar steam generation. Adv. Energy Mater. **8**, 1701028 (2018)
227. Y. Yang, X. Yang, L. Fu, M. Zou, A. Cao, Y. Du, Q. Yuan, C.H. Yan, Two-dimensional flexible bilayer janus membrane for advanced photothermal water desalination. ACS Energy Lett. **3**, 1165–1171 (2018)

228. Z. Huang, Y. Liu, S. Li, C.S. Lee, X.H. Zhang, From materials to devices: rationally designing solar steam system for advanced applications. Small Methods **6**, e2200835 (2022)
229. O. Neumann, A.D. Neumann, S. Tian, C. Thibodeaux, S. Shubhankar, J. Müller, E. Silva, A. Alabastri, S.W. Bishnoi, P. Nordlander, N.J. Halas, Combining solar steam processing and solar distillation for fully off-grid production of cellulosic bioethanol. ACS Energy Lett. **2**, 8–13 (2016)
230. E. Bellos, C. Tzivanidis, Alternative designs of parabolic trough solar collectors. Prog. Energy Combust. Sci. **71**, 81–117 (2019)
231. A.H. Elsheikh, S.W. Sharshir, M.E. Mostafa, F.A. Essa, M.K. Ahmed Ali, Applications of nanofluids in solar energy: a review of recent advances. Renew. Sustain. Energy Rev. **82**, 3483–3502 (2018)
232. A.R. Mallah, M.N. Mohd Zubir, O.A. Alawi, K.M. Salim Newaz, A.B. Mohamad Badry, Plasmonic nanofluids for high photothermal conversion efficiency in direct absorption solar collectors: fundamentals and applications. Solar Energy Mater. Solar Cells **201**, 110084 (2019)
233. O.Z. Sharaf, R.A. Taylor, E. Abu-Nada, On the colloidal and chemical stability of solar nanofluids: from nanoscale interactions to recent advances. Phys. Rep. **867**, 1–84 (2020)
234. C. Qin, K. Kang, I. Lee, B.J. Lee, Optimization of the spectral absorption coefficient of a plasmonic nanofluid for a direct absorption solar collector. Sol. Energy **169**, 231–236 (2018)
235. A.R. Mallah, S.N. Kazi, M.N.M. Zubir, A. Badarudin, Blended morphologies of plasmonic nanofluids for direct absorption applications. Appl. Energy **229**, 505–521 (2018)
236. C. Qin, J.B. Kim, B.J. Lee, Performance analysis of a direct-absorption parabolic-trough solar collector using plasmonic nanofluids. Renew. Energy **143**, 24–33 (2019)
237. O.Z. Sharaf, D.C. Kyritsis, E. Abu-Nada, Impact of nanofluids, radiation spectrum, and hydrodynamics on the performance of direct absorption solar collectors. Energy Convers. Manage. **156**, 706–722 (2018)
238. B.J. Lee, K. Park, T. Walsh, L. Xu, Radiative heat transfer analysis in plasmonic nanofluids for direct solar thermal absorption. J. Solar Energy Eng. **134** (2012)
239. J. Jeon, S. Park, B.J. Lee, Analysis on the performance of a flat-plate volumetric solar collector using blended plasmonic nanofluid. Sol. Energy **132**, 247–256 (2016)
240. M. Du, G.H. Tang, Plasmonic nanofluids based on gold nanorods/nanoellipsoids/ nanosheets for solar energy harvesting. Sol. Energy **137**, 393–400 (2016)
241. B. Fu, J. Zhang, H. Chen, H. Guo, C. Song, W. Shang, P. Tao, T. Deng, Optical nanofluids for direct absorption-based solar-thermal energy harvesting at medium-to-high temperatures. Curr. Opin. Chem. Eng. **25**, 51–56 (2019)
242. Z. Said, S. Arora, S. Farooq, L.S. Sundar, C. Li, A. Allouhi, Recent advances on improved optical, thermal, and radiative characteristics of plasmonic nanofluids: Academic insights and perspectives. Sol. Energy Mater. Sol. Cells **236**, 111504 (2022)
243. S. Santoro, A.H. Avci, A. Politano, E. Curcio, The advent of thermoplasmonic membrane distillation. Chem. Soc. Rev. **51**, 6087–6125 (2022)
244. C. Walker, E. Mitridis, T. Kreiner, H. Eghlidi, T.M. Schutzius, D. Poulikakos, Transparent metasurfaces counteracting fogging by harnessing sunlight. Nano Lett. **19**, 1595–1604 (2019)
245. X. Yin, R. Yang, G. Tan, S. Fan, Terrestrial radiative cooling: using the cold universe as a renewable and sustainable energy source. Science **370**, 786–791 (2020)
246. G. Jonsson, D. Tordera, T. Pakizeh, M. Jaysankar, V. Miljkovic, L. Tong, M.P. Jonsson, A. Dmitriev, Solar transparent radiators by optical nanoantennas. Nano Lett. **17**, 6766–6772 (2017)
247. J.C. Ndukaife, V.M. Shalaev, A. Boltasseva, Plasmonics-turning loss into gain. Science **351**, 334–335 (2016)
248. N. Maccaferri, Y. Zhao, T. Isoniemi, M. Iarossi, A. Parracino, G. Strangi, F. De Angelis, Hyperbolic meta-antennas enable full control of scattering and absorption of light. Nano Lett. **19**, 1851–1859 (2019)
249. J. Xu, J. Mandal, A.P. Raman, Broadband directional control of thermal emission. Science **372**, 393–397 (2021)

250. Y. Zhai, Y. Ma, S.N. David, D. Zhao, R. Lou, G. Tan, R. Yang, X. Yin, Scalable-manufactured randomized glass-polymer hybrid metamaterial for daytime radiative cooling. Science **355**, 1062–1066 (2017)
251. S.Y. Heo, G.J. Lee, D.H. Kim, Y.J. Kim, S. Ishii, M.S. Kim, T.J. Seok, B.J. Lee, H. Lee, Y.M. Song, A Janus emitter for passive heat release from enclosures. Sci. Adv. **6** (2020) eabb1906
252. M. Guo, L. Gao, Y. Wei, Y. Ma, Y. Jianyong, B. Ding, Solar transparent radiators based on in-plane worm-like assemblies of metal nanoparticles. Sol. Energy Mater. Sol. Cells **219**, 110796 (2021)
253. E. Mitridis, T.M. Schutzius, A. Sicher, C.U. Hail, H. Eghlidi, D. Poulikakos, Metasurfaces leveraging solar energy for icephobicity. ACS Nano **12**, 7009–7017 (2018)
254. S. Dash, J. de Ruiter, K.K. Varanasi, Photothermal trap utilizing solar illumination for ice mitigation. Sci. Adv. **4** (2018) eaat0127
255. H. Zhang, G. Zhao, S. Wu, Y. Alsaid, W. Zhao, X. Yan, L. Liu, G. Zou, J. Lv, X. He, Z. He, J. Wang, Solar anti-icing surface with enhanced condensate self-removing at extreme environmental conditions. Proc. Natl. Acad. Sci. **118**, e2100978118 (2021)
256. W. Li, C. Lin, W. Ma, Y. Li, F. Chu, B. Huang, S. Yao, Transparent selective photothermal coatings for antifogging applications. Cell Rep. Phys. Sci. **2**, 100435 (2021)
257. Y. Li, W. Ma, Y.S. Kwon, W. Li, S. Yao, B. Huang, Solar deicing nanocoatings adaptive to overhead power lines. Adv. Funct. Mater. **32** (2022)
258. P. Yao, R. Yang, Q. Sun, G.H. Tang, X. Liu, J.H. Pu, M. Du, Transparent photothermal films with high optical selectivity for anti/de-icing. Appl. Therm. Eng. **242**, 122490 (2024)
259. E. Mitridis, H. Lambley, S. Tröber, T.M. Schutzius, D. Poulikakos, Transparent photothermal metasurfaces amplifying superhydrophobicity by absorbing sunlight. ACS Nano **14**, 11712–11721 (2020)
260. Z. Chen, J. Li, Y. Zheng, Heat-mediated optical manipulation. Chem. Rev. **122**, 3122–3179 (2022)
261. H. Sipova-Jungova, D. Andren, S. Jones, M. Kall, Nanoscale inorganic motors driven by light: principles, realizations, and opportunities. Chem. Rev. **120**, 269–287 (2020)
262. J. Frueh, S. Rutkowski, T. Si, M. Gai, C. Gao, S.I. Tverdokhlebov, G. Qiu, J. Schmitt, Q. He, Y.X. Ren, J. Wang, Propulsion mechanisms of light-driven plasmonic colloidal micromotors. Adv. Photon. Res. **3** (2021)
263. I. Buttinoni, G. Volpe, F. Kummel, G. Volpe, C. Bechinger, Active Brownian motion tunable by light. J. Phys. Condens. Matter Inst. Phys. J. **24**, 284129 (2012)
264. M. Liu, T. Zentgraf, Y. Liu, G. Bartal, X. Zhang, Light-driven nanoscale plasmonic motors. Nat. Nanotechnol. **5**, 570–573 (2010)
265. J.G. Gibbs, Y.P. Zhao, Autonomously motile catalytic nanomotors by bubble propulsion. Appl. Phys. Lett. **94** (2009)
266. W. Qin, T. Peng, Y. Gao, F. Wang, X. Hu, K. Wang, J. Shi, D. Li, J. Ren, C. Fan, Catalysis-driven self-thermophoresis of janus plasmonic nanomotors. Angew. Chem. **129**, 530–533 (2016)
267. F. Meng, W. Hao, S. Yu, R. Feng, Y. Liu, F. Yu, P. Tao, W. Shang, J. Wu, C. Song, T. Deng, Vapor-enabled propulsion for plasmonic photothermal motor at the liquid/air interface. J. Am. Chem. Soc. **139**, 12362–12365 (2017)
268. X. Wang, L. Baraban, V.R. Misko, F. Nori, T. Huang, G. Cuniberti, J. Fassbender, D. Makarov, Visible light actuated efficient exclusion between plasmonic Ag/AgCl micromotors and passive beads. Small **14**, e1802537 (2018)
269. M. Franzl, S. Muinos-Landin, V. Holubec, F. Cichos, Fully steerable symmetric thermoplasmonic microswimmers. ACS Nano **15**, 3434–3440 (2021)
270. X. Wang, V. Sridhar, S. Guo, N. Talebi, A. Miguel-López, K. Hahn, P.A. van Aken, S. Sánchez, Fuel-free nanocap-like motors actuated under visible light. Adv. Funct. Mater. **28** (2018)
271. M. Xuan, Z. Wu, J. Shao, L. Dai, T. Si, Q. He, Near infrared light-powered janus mesoporous silica nanoparticle motors. J. Am. Chem. Soc. **138**, 6492–6497 (2016)
272. H. Ding, Z. Chen, C. Ponce, Y. Zheng, Optothermal rotation of micro-/nano-objects. Chem. Commun. **59**, 2208–2221 (2023)

273. L. Shao, M. Käll, Light-driven rotation of plasmonic nanomotors. Adv. Funct. Mater. **28** (2018)
274. L. Shao, Z.J. Yang, D. Andrén, P. Johansson, M. Käll, Gold nanorod rotary motors driven by resonant light scattering. ACS Nano **9**, 12542–12551 (2015)
275. L. Xin, C. Zhou, X. Duan, N. Liu, A rotary plasmonic nanoclock. Nat. Commun. **10**, 5394 (2019)
276. B. Zeng, Y. Wang, M.E. Zaytsev, C. Xia, H.J.W. Zandvliet, D. Lohse, Giant plasmonic bubbles nucleation under different ambient pressures. Phys. Rev. E **102**, 063109 (2020)
277. Y. Wang, M.E. Zaytsev, G. Lajoinie, H.L. The, J.C.T. Eijkel, A. van den Berg, M. Versluis, B.M. Weckhuysen, X. Zhang, H.J.W. Zandvliet, D. Lohse, Giant and explosive plasmonic bubbles by delayed nucleation. Proc. Natl. Acad. Sci. U.S.A. **115**, 7676–7681 (2018)
278. S. Jones, D. Andren, T.J. Antosiewicz, A. Stilgoe, H. Rubinsztein-Dunlop, M. Kall, Strong transient flows generated by thermoplasmonic bubble nucleation. ACS Nano **14**, 17468–17475 (2020)
279. W.Y. Tsai, J.S. Huang, C.B. Huang, Selective trapping or rotation of isotropic dielectric microparticles by optical near field in a plasmonic archimedes spiral. Nano Lett. **14**, 547–552 (2014)
280. M. Xuan, R. Mestre, C. Gao, C. Zhou, Q. He, S. Sánchez, Noncontinuous super-diffusive dynamics of a light-activated nanobottle motor. Angew. Chem. **130**, 6954–6958 (2018)
281. J. Feng, X. Li, T. Xu, X. Zhang, X. Du, Photothermal-driven micro/nanomotors: from structural design to potential applications. Acta Biomater. **173**, 1–35 (2024)
282. U. Das, R. Biswas, N. Mazumder, Elucidating thermal effects in plasmonic metal nanostructures: a tutorial review. Eur. Phys. J. Plus **137** (2022)
283. X. Meng, L. Liu, S. Ouyang, H. Xu, D. Wang, N. Zhao, J. Ye, Nanometals for solar-to-chemical energy conversion: from semiconductor-based photocatalysis to plasmon-mediated photocatalysis and photo-thermocatalysis. Adv. Mater. **28**, 6781–6803 (2016)
284. S. Linic, P. Christopher, D.B. Ingram, Plasmonic-metal nanostructures for efficient conversion of solar to chemical energy. Nat. Mater. **10**, 911–921 (2011)
285. Z. Zhang, C. Zhang, H. Zheng, H. Xu, Plasmon-driven catalysis on molecules and nanomaterials. Acc. Chem. Res. **52**, 2506–2515 (2019)
286. E. Cortes, W. Xie, J. Cambiasso, A.S. Jermyn, R. Sundararaman, P. Narang, S. Schlucker, S.A. Maier, Plasmonic hot electron transport drives nano-localized chemistry. Nat. Commun. **8**, 14880 (2017)
287. Y. Zhao, L. Sang, Z. Ren, Decoupling the role of thermoplasmonics effect and hot electron injection of metallic nanoparticles in photoelectrochemical system. Sol. Energy Mater. Sol. Cells **267**, 112728 (2024)
288. D. Mateo, J.L. Cerrillo, S. Durini, J. Gascon, Fundamentals and applications of photo-thermal catalysis. Chem. Soc. Rev. **50**, 2173–2210 (2021)
289. L. Mascaretti, A. Naldoni, Hot electron and thermal effects in plasmonic photocatalysis. J. Appl. Phys. **128**, 041101 (2020)
290. P. Christopher, H. Xin, S. Linic, Visible-light-enhanced catalytic oxidation reactions on plasmonic silver nanostructures. Nat. Chem. **3**, 467–472 (2011)
291. H. Huang, L. Zhang, Z. Lv, R. Long, C. Zhang, Y. Lin, K. Wei, C. Wang, L. Chen, Z.Y. Li, Q. Zhang, Y. Luo, Y. Xiong, Unraveling surface plasmon decay in core-shell nanostructures toward broadband light-driven catalytic organic synthesis. J. Am. Chem. Soc. **138**, 6822–6828 (2016)
292. P. Christopher, H. Xin, A. Marimuthu, S. Linic, Singular characteristics and unique chemical bond activation mechanisms of photocatalytic reactions on plasmonic nanostructures. Nat. Mater. **11**, 1044–1050 (2012)
293. A. Kumar, P. Choudhary, A. Kumar, P.H.C. Camargo, V. Krishnan, Recent advances in plasmonic photocatalysis based on $TiO_2$ and noble metal nanoparticles for energy conversion. Environ. Remediat. Organ. Synth. Small **18**, e2101638 (2022)
294. X. Zhang, Y.L. Chen, R.S. Liu, D.P. Tsai, Plasmonic photocatalysis, reports on progress in physics. Phys. Soc. **76**, 046401 (2013)

295. S.W. Verbruggen, $TiO_2$ photocatalysis for the degradation of pollutants in gas phase: from morphological design to plasmonic enhancement. J. Photochem. Photobiol. C **24**, 64–82 (2015)
296. J.R. Adleman, D.A. Boyd, D.G. Goodwin, D. Psaltis, Heterogenous catalysis mediated by plasmon heating. Nano Lett. **9**, 4417–4423 (2009)
297. F. Wang, C. Li, H. Chen, R. Jiang, L.D. Sun, Q. Li, J. Wang, J.C. Yu, C.H. Yan, Plasmonic harvesting of light energy for Suzuki coupling reactions. J. Am. Chem. Soc. **135**, 5588–5601 (2013)
298. J. Cui, Y. Li, L. Liu, L. Chen, J. Xu, J. Ma, G. Fang, E. Zhu, H. Wu, L. Zhao, L. Wang, Y. Huang, Near-infrared plasmonic-enhanced solar energy harvest for highly efficient photocatalytic reactions. Nano Lett. **15**, 6295–6301 (2015)
299. A. Naldoni, Z.A. Kudyshev, L. Mascaretti, S.P. Sarmah, S. Rej, J.P. Froning, O. Tomanec, J.E. Yoo, D. Wang, S. Kment, T. Montini, P. Fornasiero, V.M. Shalaev, P. Schmuki, A. Boltasseva, R. Zboril, Solar thermoplasmonic nanofurnace for high-temperature heterogeneous catalysis. Nano Lett. **20**, 3663–3672 (2020)
300. Q. Yang, Q. Xu, S.H. Yu, H.L. Jiang, Pd Nanocubes@ZIF-8: integration of plasmon-driven photothermal conversion with a metal-organic framework for efficient and selective catalysis. Angew. Chem. **128**, 3749–3753 (2016)
301. W.H. Hung, M. Aykol, D. Valley, W. Hou, S.B. Cronin, Plasmon resonant enhancement of carbon monoxide catalysis. Nano Lett. **10**, 1314–1318 (2010)
302. X. Huang, S. Tang, X. Mu, Y. Dai, G. Chen, Z. Zhou, F. Ruan, Z. Yang, N. Zheng, Freestanding palladium nanosheets with plasmonic and catalytic properties. Nat. Nanotechnol. **6**, 28–32 (2011)
303. D.F. Swearer, H. Zhao, L. Zhou, C. Zhang, H. Robatjazi, J.M.P. Martirez, C.M. Krauter, S. Yazdi, M.J. McClain, E. Ringe, E.A. Carter, P. Nordlander, N.J. Halas, Heterometallic antenna−reactor complexes for photocatalysis. Proc. Natl. Acad. Sci. **113**, 8916–8920 (2016)
304. U. Aslam, S. Chavez, S. Linic, Controlling energy flow in multimetallic nanostructures for plasmonic catalysis. Nat. Nanotechnol. **12**, 1000–1005 (2017)
305. A. Marimuthu, J. Zhang, S. Linic, Tuning selectivity in propylene epoxidation by plasmon mediated photo-switching of Cu oxidation state. Science **339**, 1590–1593 (2013)
306. L. Zhou, J.M.P. Martirez, J. Finzel, C. Zhang, D.F. Swearer, S. Tian, H. Robatjazi, M. Lou, L. Dong, L. Henderson, P. Christopher, E.A. Carter, P. Nordlander, N.J. Halas, Light-driven methane dry reforming with single atomic site antenna-reactor plasmonic photocatalysts. Nat. Energy **5**, 61–70 (2020)
307. C. Zhan, Q.X. Wang, J. Yi, L. Chen, D.Y. Wu, Y. Wang, Z.X. Xie, M. Moskovits, Z.Q. Tian, Plasmonic nanoreactors regulating selective oxidation by energetic electrons and nanoconfined thermal fields. Sci. Adv. **7** (2021) eabf0962
308. L. Zhou, D.F. Swearer, C. Zhang, H. Robatjazi, H. Zhao, L. Henderson, L. Dong, P. Christopher, E.A. Carter, P. Nordlander, N.J. Halas, Quantifying hot carrier and thermal contributions in plasmonic photocatalysis. Science **362**, 69–72 (2018)
309. C. Vázquez-Vázquez, B. Vaz, V. Giannini, M. Pérez-Lorenzo, R.A. Alvarez-Puebla, M.A. Correa-Duarte, Nanoreactors for simultaneous remote thermal activation and optical monitoring of chemical reactions. J. Am. Chem. Soc. **135**, 13616–13619 (2013)
310. J.H. Kim, D. Hansora, P. Sharma, J.W. Jang, J.S. Lee, Toward practical solar hydrogen production-an artificial photosynthetic leaf-to-farm challenge. Chem. Soc. Rev. **48**, 1908–1971 (2019)
311. J. Zhang, X. Jin, P.I. Morales-Guzman, X. Yu, H. Liu, H. Zhang, L. Razzari, J.P. Claverie, Engineering the absorption and field enhancement properties of Au-$TiO_2$ nanohybrids via whispering gallery mode resonances for photocatalytic water splitting. ACS Nano **10**, 4496–4503 (2016)
312. S.S. Yi, X.B. Zhang, B.R. Wulan, J.M. Yan, Q. Jiang, Non-noble metals applied to solar water splitting. Energy Environ. Sci. **11**, 3128–3156 (2018)
313. P. Zhang, T. Wang, J. Gong, Mechanistic understanding of the plasmonic enhancement for solar water splitting. Adv. Mater. **27**, 5328–5342 (2015)

314. S.C. Warren, E. Thimsen, Plasmonic solar water splitting. Energy Environ. Sci. **5**, 5133–5146 (2012)
315. H. Sheng, J. Wang, J. Huang, Z. Li, G. Ren, L. Zhang, L. Yu, M. Zhao, X. Li, G. Li, N. Wang, C. Shen, G. Lu, Strong synergy between gold nanoparticles and cobalt porphyrin induces highly efficient photocatalytic hydrogen evolution. Nat. Commun. **14**, 1528 (2023)
316. M. Herran, S. Juergensen, M. Kessens, D. Hoeing, A. Köppen, A. Sousa-Castillo, W.J. Parak, H. Lange, S. Reich, F. Schulz, E. Cortés, Plasmonic bimetallic two-dimensional supercrystals for $H_2$ generation. Nat. Catal. **6**, 1205–1214 (2023)
317. Y. Zhang, W. Guo, Y. Zhang, W.D. Wei, Plasmonic photoelectrochemistry: in view of hot carriers. Adv. Mater. 2006654 (2021)
318. G. Liu, K. Du, J. Xu, G. Chen, M. Gu, C. Yang, K. Wang, H. Jakobsen, Plasmon-dominated photoelectrodes for solar water splitting. J. Mater. Chem. A **5**, 4233–4253 (2017)
319. D.V. Dao, T.T.D. Nguyen, P. Uthirakumar, Y.-H. Cho, G.C. Kim, J.K. Yang, D.T. Tran, T.D. Le, H. Choi, H.Y. Kim, Y.T. Yu, I.H. Lee, Insightful understanding of hot-carrier generation and transfer in plasmonic Au@$CeO_2$ core–shell photocatalysts for light-driven hydrogen evolution improvement. Appl. Catal. B **286**, 119947 (2021)
320. J.L. Yang, Y.L. He, H. Ren, H.L. Zhong, J.S. Lin, W.-M. Yang, M.D. Li, Z.L. Yang, H. Zhang, Z.Q. Tian, J.F. Li, Boosting photocatalytic hydrogen evolution reaction using dual plasmonic antennas. ACS Catal. **11**, 5047–5053 (2021)
321. Q. Wang, T. Hisatomi, Q. Jia, H. Tokudome, M. Zhong, C. Wang, Z. Pan, T. Takata, M. Nakabayashi, N. Shibata, Y. Li, I.D. Sharp, A. Kudo, T. Yamada, K. Domen, Scalable water splitting on particulate photocatalyst sheets with a solar-to-hydrogen energy conversion efficiency exceeding 1. Nat. Mater. **15**, 611–615 (2016)
322. S. Wang, Y. Gao, S. Miao, T. Liu, L. Mu, R. Li, F. Fan, C. Li, Positioning the water oxidation reaction sites in plasmonic photocatalysts. J. Am. Chem. Soc. **139**, 11771–11778 (2017)
323. Z.W. Seh, S. Liu, M. Low, S.Y. Zhang, Z. Liu, A. Mlayah, M.Y. Han, Janus Au-$TiO_2$ photocatalysts with strong localization of plasmonic near-fields for efficient visible-light hydrogen generation. Adv. Mater. **24**, 2310–2314 (2012)
324. K. Song, H. Lee, M. Lee, J.Y. Park, Plasmonic hot hole-driven water splitting on Au nanoprisms/P-Type GaN. ACS Energy Lett. 1333–1339 (2021)
325. Z. Zhang, L. Zhang, M.N. Hedhili, H. Zhang, P. Wang, Plasmonic gold nanocrystals coupled with photonic crystal seamlessly on $TiO_2$ nanotube photoelectrodes for efficient visible light photoelectrochemical water splitting. Nano Lett. **13**, 14–20 (2013)
326. X. Shi, K. Ueno, T. Oshikiri, Q. Sun, K. Sasaki, H. Misawa, Enhanced water splitting under modal strong coupling conditions. Nat. Nanotechnol. **13**, 953–958 (2018)
327. R. Peng, X. Ma, Z.D. Hood, A. Boulesbaa, A.A. Puretzky, J. Tong, Z. Wu, Synergizing plasmonic Au nanocages with 2D $MoS_2$ nanosheets for significant enhancement in photocatalytic hydrogen evolution. J. Mater. Chem. A **11**, 16714–16723 (2023)
328. P. Verma, Y. Kuwahara, K. Mori, R. Watanabe, C. Fukuhara, H. Yamashita, Plasmonic heterojunction photocatalysts for hydrogen generation: a mini-review. Energy Fuels **37**, 17652–17666 (2023)
329. G. Yu, J. Qian, P. Zhang, B. Zhang, W. Zhang, W. Yan, G. Liu, Collective excitation of plasmon-coupled Au-nanochain boosts photocatalytic hydrogen evolution of semiconductor. Nat. Commun. **10** (2019)
330. N. Keller, J. Ivanez, J. Highfield, A.M. Ruppert, Photo-/thermal synergies in heterogeneous catalysis: towards low-temperature (solar-driven) processing for sustainable energy and chemicals. Appl. Catal. B **296**, 120320 (2021)
331. K. Czelej, J.C. Colmenares, K. Jabłczyńska, K. Ćwieka, Ł. Werner, L. Gradoń, Sustainable hydrogen production by plasmonic thermophotocatalysis. Catal. Today (2021)
332. S. Guo, X. Li, J. Li, B. Wei, Boosting photocatalytic hydrogen production from water by photothermally induced biphase systems. Nat. Commun. **12** (2021)
333. L. Ma, B. Luo, J. Geng, Z. Huang, L. Guo, Efficient photothermocatalytic hydrogen production performance over a graphene-titanium dioxide hybrid nanomaterial. Int. J. Hydrogen Energy **46**, 2871–2877 (2021)

334. Q. Li, G. Lu, Significant effect of pressure on the $H_2$ releasing from photothermal-catalytic water steam splitting over $TiSi_2$ and $Pt/TiO_2$. Catal. Lett. **125**, 376–379 (2008)
335. T.G.U. Ghobadi, A. Ghobadi, E. Ozbay, F. Karadas, Strategies for plasmonic hot-electron-driven photoelectrochemical water splitting. Chemphotochem **2**, 161–182 (2018)
336. Z.J. Wang, H. Song, H. Liu, J. Ye, Coupling of solar energy and thermal energy for carbon dioxide reduction: status and prospects. Angew. Chem. **59**, 8016–8035 (2020)
337. J. He, C. Janaky, Recent advances in solar-driven carbon dioxide conversion: expectations versus reality. ACS Energy Lett. **5**, 1996–2014 (2020)
338. M. Dhiman, Plasmonic nanocatalysis for solar energy harvesting and sustainable chemistry. J. Mater. Chem. A **8**, 10074–10095 (2020)
339. T. Inoue, A. Fujishima, S. Konishi, K. Honda, Photoelectrocatalytic reduction of carbon dioxide in aqueous suspensions of semiconductor powders. Nature **277**, 637–638 (1979)
340. S. Yu, A.J. Wilson, G. Kumari, X. Zhang, P.K. Jain, Opportunities and challenges of solar-energy-driven carbon dioxide to fuel conversion with plasmonic catalysts. ACS Energy Lett. **2**, 2058–2070 (2017)
341. U. Ulmer, T. Dingle, P.N. Duchesne, R.H. Morris, A. Tavasoli, T. Wood, G.A. Ozin, Fundamentals and applications of photocatalytic $CO_2$ methanation. Nat. Commun. **10**, 3169 (2019)
342. X. Zhang, X. Li, M.E. Reish, D. Zhang, N.Q. Su, Y. Gutierrez, F. Moreno, W. Yang, H.O. Everitt, J. Liu, Plasmon-enhanced catalysis: distinguishing thermal and nonthermal effects. Nano Lett. **18**, 1714–1723 (2018)
343. Q. Kang, T. Wang, P. Li, L. Liu, K. Chang, M. Li, J. Ye, Photocatalytic reduction of carbon dioxide by hydrous hydrazine over Au-Cu alloy nanoparticles supported on $SrTiO_3/TiO_2$ coaxial nanotube arrays. Angew. Chem. **54**, 841–845 (2015)
344. M. Dhiman, A. Maity, A. Das, R. Belgamwar, B. Chalke, Y. Lee, K. Sim, J.M. Nam, V. Polshettiwar, Plasmonic colloidosomes of black gold for solar energy harvesting and hotspots directed catalysis for $CO_2$ to fuel conversion. Chem. Sci. **10**, 6594–6603 (2019)
345. X. Zhang, X. Li, D. Zhang, N.Q. Su, W. Yang, H.O. Everitt, J. Liu, Product selectivity in plasmonic photocatalysis for carbon dioxide hydrogenation. Nat. Commun. **8**, 14542 (2017)
346. J.S. DuChene, G. Tagliabue, A.J. Welch, W.H. Cheng, H.A. Atwater, Hot hole collection and photoelectrochemical $CO_2$ reduction with plasmonic Au/p-GaN photocathodes. Nano Lett. **18**, 2545–2550 (2018)
347. G. Kumari, X. Zhang, D. Devasia, J. Heo, P.K. Jain, Watching visible light-driven $CO_2$ reduction on a plasmonic nanoparticle catalyst. ACS Nano **12**, 8330–8340 (2018)
348. W.N. Wang, W.J. An, B. Ramalingam, S. Mukherjee, D.M. Niedzwiedzki, S. Gangopadhyay, P. Biswas, Size and structure matter: enhanced $CO_2$ photoreduction efficiency by size-resolved ultrafine Pt nanoparticles on $TiO_2$ single crystals. J. Am. Chem. Soc. **134**, 11276–11281 (2012)
349. H. Liu, X. Meng, T.D. Dao, H. Zhang, P. Li, K. Chang, T. Wang, M. Li, T. Nagao, J. Ye, Conversion of carbon dioxide by methane reforming under visible-light irradiation: surface-plasmon-mediated nonpolar molecule activation. Angew. Chem. Int. Ed. **54**, 11545–11549 (2015)
350. S. Neatu, J.A. Macia-Agullo, P. Concepcion, H. Garcia, Gold-copper nanoalloys supported on $TiO_2$ as photocatalysts for $CO_2$ reduction by water. J. Am. Chem. Soc. **136**, 15969–15976 (2014)
351. R. Li, W.H. Cheng, M.H. Richter, J.S. DuChene, W. Tian, C. Li, H.A. Atwater, Unassisted highly selective gas-phase $CO_2$ reduction with a plasmonic Au/p-GaN photocatalyst using $H_2O$ as an electron donor. ACS Energy Lett. **6**, 1849–1856 (2021)
352. J. Zhao, B. Liu, L. Meng, S. He, R. Yuan, Y. Hou, Z. Ding, H. Lin, Z. Zhang, X. Wang, J. Long, Plasmonic control of solar-driven $CO_2$ conversion at the metal/ZnO interfaces. Appl. Catal. B **256**, 117823 (2019)
353. X. Meng, T. Wang, L. Liu, S. Ouyang, P. Li, H. Hu, T. Kako, H. Iwai, A. Tanaka, J. Ye, Photothermal conversion of $CO_2$ into $CH_4$ with $H_2$ over Group VIII nanocatalysts: an alternative approach for solar fuel production. Angew. Chem. **53**, 11478–11482 (2014)

354. M. Ghoussoub, M. Xia, P.N. Duchesne, D. Segal, G. Ozin, Principles of photothermal gas-phase heterogeneous $CO_2$ catalysis. Energy Environ. Sci. **12**, 1122–1142 (2019)
355. Y. Li, J. Hao, H. Song, F. Zhang, X. Bai, X. Meng, H. Zhang, S. Wang, Y. Hu, J. Ye, Selective light absorber-assisted single nickel atom catalysts for ambient sunlight-driven $CO_2$ methanation. Nat. Commun. **10**, 2359 (2019)
356. J. Jia, H. Wang, Z. Lu, P.G. O'Brien, M. Ghoussoub, P. Duchesne, Z. Zheng, P. Li, Q. Qiao, L. Wang, A. Gu, A.A. Jelle, Y. Dong, Q. Wang, K.K. Ghuman, T. Wood, C. Qian, Y. Shao, C. Qiu, M. Ye, Y. Zhu, Z.H. Lu, P. Zhang, A.S. Helmy, C.V. Singh, N.P. Kherani, D.D. Perovic, G.A. Ozin, Photothermal catalyst engineering: hydrogenation of gaseous $CO_2$ with high activity and tailored selectivity. Adv. Sci. **4**, 1700252 (2017)
357. M. Cai, Z. Wu, Z. Li, L. Wang, W. Sun, A.A. Tountas, C. Li, S. Wang, K. Feng, A.-B. Xu, S. Tang, A. Tavasoli, M. Peng, W. Liu, A.S. Helmy, L. He, G.A. Ozin, X. Zhang, Greenhouse-inspired supra-photothermal $CO_2$ catalysis. Nat. Energy **6**, 807–814 (2021)
358. M.L. Brongersma, N.J. Halas, P. Nordlander, Plasmon-induced hot carrier science and technology. Nat. Nanotechnol. **10**, 25–34 (2015)
359. E.C. Edited by Pedro H.C. Camargo, Plasmonic Catalysis from Fundamentals to Applications, © 2021 WILEY-VCH GmbH (2021)
360. B.M. Comer, P. Fuentes, C.O. Dimkpa, Y.H. Liu, C.A. Fernandez, P. Arora, M. Realff, U. Singh, M.C. Hatzell, A.J. Medford, Prospects and challenges for solar fertilizers. Joule **3**, 1578–1605 (2019)
361. B. Puértolas, M. Comesaña-Hermo, L.V. Besteiro, M. Vázquez-González, M.A. Correa-Duarte, challenges and opportunities for renewable ammonia production via plasmon-assisted photocatalysis. Adv. Energy Mater. **12** (2022)
362. J.G. Chen, R.M. Crooks, L.C. Seefeldt, K.L. Bren, R.M. Bullock, M.Y. Darensbourg, P.L. Holland, B. Hoffman, M.J. Janik, A.K. Jones, M.G. Kanatzidis, P. King, K.M. Lancaster, S.V. Lymar, P. Pfromm, W.F. Schneider, R.R. Schrock, Beyond fossil fuel-driven nitrogen transformations. Science **360** (2018)
363. R. Schlogl, Catalytic synthesis of ammonia-a "never-ending story"? Angew. Chem. **42**, 2004–2008 (2003)
364. S.D. Minteer, P. Christopher, S. Linic, Recent developments in nitrogen reduction catalysts: a virtual issue. ACS Energy Lett. **4**, 163–166 (2018)
365. L. Wang, M. Xia, H. Wang, K. Huang, C. Qian, C.T. Maravelias, G.A. Ozin, Greening ammonia toward the solar ammonia refinery. Joule **2**, 1055–1074 (2018)
366. R. de Richter, T. Ming, P. Davies, W. Liu, S. Caillol, Removal of non-$CO_2$ greenhouse gases by large-scale atmospheric solar photocatalysis.Progr. Energy Combustion Sci. **60**, 68–96 (2017)
367. J. Yang, Y. Guo, W. Lu, R. Jiang, J. Wang, Emerging applications of plasmons in driving $CO_2$ reduction and $N_2$ fixation. Adv. Mater. **30**, e1802227 (2018)
368. E. Cortes, L.V. Besteiro, A. Alabastri, A. Baldi, G. Tagliabue, A. Demetriadou, P. Narang, Challenges in plasmonic catalysis. ACS Nano (2020)
369. C. Li, T. Wang, Z.J. Zhao, W. Yang, J.F. Li, A. Li, Z. Yang, G.A. Ozin, J. Gong, Promoted fixation of molecular nitrogen with surface oxygen vacancies on plasmon-enhanced $TiO_2$ photoelectrodes. Angew. Chem. Int. Ed. **57**, 5278–5282 (2018)
370. T. Oshikiri, K. Ueno, H. Misawa, Plasmon-induced ammonia synthesis through nitrogen photofixation with visible light irradiation. Angew. Chem. **53**, 9802–9805 (2014)
371. T. Oshikiri, K. Ueno, H. Misawa, Selective dinitrogen conversion to ammonia using water and visible light through plasmon-induced charge separation. Angew. Chem. **55**, 3942–3946 (2016)
372. M. Ali, F. Zhou, K. Chen, C. Kotzur, C. Xiao, L. Bourgeois, X. Zhang, D.R. MacFarlane, Nanostructured photoelectrochemical solar cell for nitrogen reduction using plasmon-enhanced black silicon. Nat. Commun. **7**, 11335 (2016)
373. X. Li, X. Zhang, H.O. Everitt, J. Liu, Light-induced thermal gradients in ruthenium catalysts significantly enhance ammonia production. Nano Lett. **19**, 1706–1711 (2019)

374. C. Mao, H. Li, H. Gu, J. Wang, Y. Zou, G. Qi, J. Xu, F. Deng, W. Shen, J. Li, S. Liu, J. Zhao, L. Zhang, Beyond the thermal equilibrium limit of ammonia synthesis with dual temperature zone catalyst powered by solar light. Chem **5**, 2702–2717 (2019)
375. C. Mao, L. Yu, J. Li, J. Zhao, L. Zhang, Energy-confined solar thermal ammonia synthesis with $K/Ru/TiO_{2-x}H_x$. Appl. Catal. B **224**, 612–620 (2018)
376. S. Zhang, D. Chen, P. Chen, R. Zhang, Y. Hou, Y. Guo, P. Li, X. Liang, T. Xing, J. Chen, Y. Zhao, Z. Huang, D. Lei, C. Zhi, Concurrent mechanisms of hot electrons and interfacial water molecule ordering in plasmon-enhanced nitrogen fixation. Adv. Mater. e2310776 (2024)
377. Y. Yang, H. Jia, N. Hu, M. Zhao, J. Li, W. Ni, C.Y. Zhang, Construction of gold/rhodium freestanding superstructures as antenna-reactor photocatalysts for plasmon-driven nitrogen fixation. J. Am. Chem. Soc. **146**, 7734–7742 (2024)
378. S.K. Boong, C. Chong, J.K. Lee, Z.Z. Ang, H. Li, H.K. Lee, Superlattice-based plasmonic catalysis: concentrating light at the nanoscale to drive efficient nitrogen-to-ammonia fixation at ambient conditions. Angew. Chem. **62**, e202216562 (2023)
379. X. Li, B. Sun, Q. Chen, H.K. Lee, B. Shi, H. Ren, H. Li, Z. Ma, M. Fu, Integrating photothermal and plasmonic catalysis induced by near-infrared light for efficient reduction of 4-nitrophenol. J. Colloid Interface Sci. **660**, 726–734 (2024)
380. Y. Feng, L. Dai, Z. Wang, Y. Peng, E. Duan, Y. Liu, L. Jing, X. Wang, A. Rastegarpanah, H. Dai, J. Deng, Photothermal synergistic effect of $Pt_1/CuO\text{-}CeO_2$ single-atom catalysts significantly improving toluene removal. Environ. Sci. Technol. **56**, 8722–8732 (2022)
381. Z. Zhang, A. Zhao, F. Wang, J. Ren, X. Qu, Design of a plasmonic micromotor for enhanced photo-remediation of polluted anaerobic stagnant waters. Chem. Commun. **52**, 5550–5553 (2016)
382. Y. Deng, L. Tang, C. Feng, G. Zeng, J. Wang, Y. Zhou, Y. Liu, B. Peng, H. Feng, Construction of plasmonic Ag modified phosphorous-doped ultrathin $g\text{-}C_3N_4$ nanosheets/$BiVO_4$ photocatalyst with enhanced visible-near-infrared response ability for ciprofloxacin degradation. J. Hazard. Mater. **344**, 758–769 (2018)
383. J.W. Hong, D.H. Wi, S.U. Lee, S.W. Han, Metal-semiconductor heteronanocrystals with desired configurations for plasmonic photocatalysis. J. Am. Chem. Soc. **138**, 15766–15773 (2016)
384. M. Wang, M. Ye, J. Iocozzia, C. Lin, Z. Lin, Plasmon-mediated solar energy conversion via photocatalysis in noble metal/semiconductor composites. Adv. Sci. **3**, 1600024 (2016)
385. Z. Bian, T. Tachikawa, P. Zhang, M. Fujitsuka, T. Majima, $Au/TiO_2$ superstructure-based plasmonic photocatalysts exhibiting efficient charge separation and unprecedented activity. J. Am. Chem. Soc. **136**, 458–465 (2014)
386. C. Chen, L. Zhou, J. Yu, Y. Wang, S. Nie, S. Zhu, J. Zhu, Dual functional asymmetric plasmonic structures for solar water purification and pollution detection. Nano Energy **51**, 451–456 (2018)
387. S.W. Verbruggen, M. Keulemans, B. Goris, N. Blommaerts, S. Bals, J.A. Martens, S. Lenaerts, Plasmonic 'rainbow' photocatalyst with broadband solar light response for environmental applications. Appl. Catal. B **188**, 147–153 (2016)
388. J. Huang, Y. He, L. Wang, Y. Huang, B. Jiang, Bifunctional $Au@TiO_2$ core–shell nanoparticle films for clean water generation by photocatalysis and solar evaporation. Energy Convers. Manage. **132**, 452–459 (2017)
389. K.Y. Tang, J.X. Chen, E.D.R. Legaspi, C. Owh, M. Lin, I.S.Y. Tee, D. Kai, X.J. Loh, Z. Li, M.D. Regulacio, E. Ye, Gold-decorated $TiO_2$ nanofibrous hybrid for improved solar-driven photocatalytic pollutant degradation. Chemosphere **265**, 129114 (2021)
390. H.A. Ghaly, A.S. El-Kalliny, T.A. Gad-Allah, N.E.A. Abd El-Sattar, E.R. Souaya, Stable plasmonic Ag/AgCl-polyaniline photoactive composite for degradation of organic contaminants under solar light. RSC Adv. **7**, 12726–12736 (2017)
391. C. Yu, W. Zhou, L. Zhu, G. Li, K. Yang, R. Jin, Integrating plasmonic Au nanorods with dendritic like $\alpha\text{-}Bi_2O_3/Bi_2O_2CO_3$ heterostructures for superior visible-light-driven photocatalysis. Appl. Catal. B **184**, 1–11 (2016)
392. G. Baffou, R. Quidant, Nanoplasmonics for chemistry. Chem. Soc. Rev. **43**, 3898–3907 (2014)

393. C. Zhan, B.W. Liu, Y.F. Huang, S. Hu, B. Ren, M. Moskovits, Z.Q. Tian, Disentangling charge carrier from photothermal effects in plasmonic metal nanostructures. Nat. Commun. **10**, 2671 (2019)
394. X. Li, H.O. Everitt, J. Liu, Synergy between thermal and nonthermal effects in plasmonic photocatalysis. Nano Res. **13**, 1268–1280 (2020)
395. H. Wei, S.K. Loeb, N.J. Halas, J.H. Kim, Plasmon-enabled degradation of organic micropollutants in water by visible-light illumination of Janus gold nanorods. Proc. Natl. Acad. Sci. **117**, 15473–15481 (2020)
396. S.C. Cai, J.J. Li, E.Q. Yu, X. Chen, J. Chen, H.P. Jia, Strong photothermal effect of plasmonic Pt nanoparticles for efficient degradation of volatile organic compounds under solar light irradiation. ACS Appl. Nano Mater. **1**, 6368–6377 (2018)
397. X. Liu, J. Iocozzia, Y. Wang, X. Cui, Y. Chen, S. Zhao, Z. Li, Z. Lin, Noble metal–metal oxide nanohybrids with tailored nanostructures for efficient solar energy conversion, photocatalysis and environmental remediation. Energy Environ. Sci. **10**, 402–434 (2017)
398. I. Jeon, E.C. Ryberg, P.J.J. Alvarez, J.H. Kim, Technology assessment of solar disinfection for drinking water treatment. Nat. Sustain. **5**, 801–808 (2022)
399. P.J.J. Alvarez, C.K. Chan, M. Elimelech, N.J. Halas, D. Villagran, Emerging opportunities for nanotechnology to enhance water security. Nat. Nanotechnol. **13**, 634–641 (2018)
400. J. Gamage McEvoy, Z. Zhang, Antimicrobial and photocatalytic disinfection mechanisms in silver-modified photocatalysts under dark and light conditions. J. Photochem. Photobiol. C: Photochem. Rev. **19**, 62–75 (2014)
401. L. Wang, Y. Feng, K. Wang, G. Liu, Solar water sterilization enabled by photothermal nanomaterials. Nano Energy **87**, 106158 (2021)
402. O. Neumann, C. Feronti, A.D. Neumann, A. Dong, K. Schell, B. Lu, E. Kim, M. Quinn, S. Thompson, N. Grady, P. Nordlander, M. Oden, N.J. Halas, Compact solar autoclave based on steam generation using broadband light-harvesting nanoparticles. Proc. Natl. Acad. Sci. U.S.A. **110**, 11677–11681 (2013)
403. A. Kulkarni, A. Kapley, R.S. Dhodapkar, P. Nagababu, S. Rayalu, Plasmonics driven engineered pasteurizers for solar water disinfection (SWADIS). J. Hazard. Mater. **369**, 474–482 (2019)
404. F. Brasili, A. Capocefalo, D. Palmieri, F. Capitani, E. Chiessi, G. Paradossi, F. Bordi, F. Domenici, Assembling patchy plasmonic nanoparticles with aggregation-dependent antibacterial activity. J. Colloid Interface Sci. **580**, 419–428 (2020)
405. J. Li, M. Du, G. Lv, L. Zhou, X. Li, L. Bertoluzzi, C. Liu, S. Zhu, J. Zhu, Interfacial solar steam generation enables fast-responsive, energy-efficient, and low-cost off-grid sterilization. Adv. Mater. **30**, e1805159 (2018)
406. S.K. Loeb, J. Kim, C. Jiang, L.S. Early, H. Wei, Q. Li, J.H. Kim, Nanoparticle enhanced interfacial solar photothermal water disinfection demonstrated in 3-D printed flow-through reactors. Environ. Sci. Technol. **53**, 7621–7631 (2019)
407. D. Krueger, A. Graves, Y. Chu, J. Pan, S.E. Lee, Y. Park, Integrated plasmonic gold nanoparticle dimer array for sustainable solar water disinfection. ACS Appl. Nano Mater. **6**, 5568–5577 (2023)
408. L. Zhao, B. Bhatia, L. Zhang, E. Strobach, A. Leroy, M.K. Yadav, S. Yang, T.A. Cooper, L.A. Weinstein, A. Modi, S.B. Kedare, G. Chen, E.N. Wang, A passive high-temperature high-pressure solar steam generator for medical sterilization. Joule **4**, 2733–2745 (2020)
409. S.B. Borrelle, J. Ringma, K.L. Law, C.C. Monnahan, L. Lebreton, A. McGivern, E. Murphy, J. Jambeck, G.H. Leonard, M.A. Hilleary, M. Eriksen, H.P. Possingham, H. De Frond, L.R. Gerber, B. Polidoro, A. Tahir, M. Bernard, N. Mallos, M. Barnes, C.M. Rochman, Predicted growth in plastic waste exceeds efforts to mitigate plastic pollution. Science **369**, 1515–1518 (2020)
410. L. Cabernard, S. Pfister, C. Oberschelp, S. Hellweg, Growing environmental footprint of plastics driven by coal combustion. Nature Sustainability **5**, 139–148 (2021)
411. C. Wilcox, E. Van Sebille, B.D. Hardesty, Threat of plastic pollution to seabirds is global, pervasive, and increasing. Proc. Natl. Acad. Sci. U.S.A. **112**, 11899–11904 (2015)

412. A. Stubbins, K.L. Law, S.E. Munoz, T.S. Bianchi, L. Zhu, Plastics in the Earth system. Science **373**, 51–55 (2021)
413. J. Yates, M. Deeney, H.B. Rolker, H. White, S. Kalamatianou, S. Kadiyala, A systematic scoping review of environmental, food security and health impacts of food system plastics. Nat. food **2**, 80–87 (2021)
414. G.J. Mahler, M.B. Esch, E. Tako, T.L. Southard, S.D. Archer, R.P. Glahn, M.L. Shuler, Oral exposure to polystyrene nanoparticles affects iron absorption. Nat. Nanotechnol. **7**, 264–271 (2012)
415. P. Wang, Z. Huang, S. Chen, M. Jing, Z. Ge, J. Chen, S. Yang, J. Chen, Y. Fang, Sustainable removal of nano/microplastics in water by solar energy. Chem. Eng. J. **428**, 131196 (2022)
416. X. Dong, X. Liu, Q. Hou, Z. Wang, From natural environment to animal tissues: a review of microplastics (nanoplastics) translocation and hazards studies. Sci. Total. Environ. **855**, 158686 (2023)
417. Y. Pan, S.H. Gao, C. Ge, Q. Gao, S. Huang, Y. Kang, G. Luo, Z. Zhang, L. Fan, Y. Zhu, A.J. Wang, Removing microplastics from aquatic environments: a critical review. Environ. Sci. Ecotechnol. **13**, 100222 (2023)
418. M. Padervand, E. Lichtfouse, D. Robert, C. Wang, Removal of microplastics from the environment. A review. Environ. Chem. Lett. **18**, 807–828 (2020)
419. Y. Zhang, H. Jiang, K. Bian, H. Wang, C. Wang, Is froth flotation a potential scheme for microplastics removal? Anal. Flotat. Kinet. Surf. Characterist. Sci. Total Environ. **792**, 148345 (2021)
420. K.H. Shosuke Yoshida, T. Takehana, I. Taniguchi, H. Yamaji, Y. Maeda, K. Toyohara, K. Miyamoto,Y. Kimura, K. Oda, A bacterium that degrades and assimilates poly(ethylene terephthalate). Science **351**, 1196–1119 (2016)
421. M. Enfrin, J. Lee, P. Le-Clech, L.F. Dumée, Kinetic and mechanistic aspects of ultrafiltration membrane fouling by nano- and microplastics. J. Membr. Sci. **601**, 117890 (2020)
422. S. Moon, L.M.A. Martin, S. Kim, Q. Zhang, R. Zhang, W. Xu, T. Luo, Direct observation and identification of nanoplastics in ocean water. Sci. Adv. **10** (2024) eadh1675
423. B.J. Roxworthy, A.M. Bhuiya, S.P. Vanka, K.C. Toussaint Jr., Understanding and controlling plasmon-induced convection. Nat. Commun. **5**, 3173 (2014)
424. C. Zhao, Y. Xie, Z. Mao, Y. Zhao, J. Rufo, S. Yang, F. Guo, J.D. Mai, T.J. Huang, Theory and experiment on particle trapping and manipulation via optothermally generated bubbles. Lab Chip **14**, 384–391 (2014)
425. H. Ye, Y. Wang, X. Liu, D. Xu, H. Yuan, H. Sun, S. Wang, X. Ma, Magnetically steerable iron oxides-manganese dioxide core-shell micromotors for organic and microplastic removals. J. Colloid Interface Sci. **588**, 510–521 (2021)
426. K.M. Wyss, J.T. Li, P.A. Advincula, K.V. Bets, W. Chen, L. Eddy, K.J. Silva, J.L. Beckham, J. Chen, W. Meng, B. Deng, S. Nagarajaiah, B.I. Yakobson, J.M. Tour, Upcycling of waste plastic into hybrid carbon nanomaterials. Adv. Mater. **35**, e2209621 (2023)
427. K. Wang, E. Schonbrun, P. Steinvurzel, K.B. Crozier, Trapping and rotating nanoparticles using a plasmonic nano-tweezer with an integrated heat sink. Nat. Commun. **2**, 469 (2011)
428. J.A. Darr, J. Zhang, N.M. Makwana, X. Weng, Continuous hydrothermal synthesis of inorganic nanoparticles: applications and future directions. Chem. Rev. **117**, 11125–11238 (2017)
429. M. Ceballos, S. Funes-Hernando, G. Zampini, M. Cedrún-Morales, J.M. Vila-Fungueiriño, B. Pelaz, P. del Pino, Seeded-growth of PCN-224 onto plasmonic nanoparticles: photoactive microporous nanocarriers. Small Structur. **5** (2024)
430. Q.N. Nguyen, C. Wang, Y. Shang, A. Janssen, Y. Xia, Colloidal synthesis of metal nanocrystals: from asymmetrical growth to symmetry breaking. Chem. Rev. **123**, 3693–3760 (2023)
431. J. Lee, J. Yang, S.G. Kwon, T. Hyeon, Nonclassical nucleation and growth of inorganic nanoparticles. Nat. Rev. Mater. **1** (2016)
432. M. Strach, V. Mantella, J.R. Pankhurst, P. Iyengar, A. Loiudice, S. Das, C. Corminboeuf, W. van Beek, R. Buonsanti, Insights into reaction intermediates to predict synthetic pathways for shape-controlled metal nanocrystals. J. Am. Chem. Soc. **141**, 16312–16322 (2019)

433. A. Trovarelli, J. Llorca, Ceria catalysts at nanoscale: how do crystal shapes shape catalysis? ACS Catal. **7**, 4716–4735 (2017)
434. A. Shinde, D. Huang, M. Saldivar, H. Xu, M. Zeng, U. Okeibunor, L. Wang, C. Mejia, P. Tin, S. George, L. Zhang, Z. Cheng, Growth of colloidal nanoplate liquid crystals using temperature gradients. ACS Nano **13**, 12461–12469 (2019)
435. B. Huang, L. Miao, J. Li, Z. Xie, Y. Wang, J. Chai, Y. Zhai, Identification of plasmon-driven nanoparticle-coalescence-dominated growth of gold nanoplates through nanopore sensing. Nat. Commun. **13**, 1402 (2022)
436. A. Kumar, K.W. Jeon, N. Kumari, I.S. Lee, Spatially confined formation and transformation of nanocrystals within nanometer-sized reaction media. Acc. Chem. Res. **51**, 2867–2879 (2018)
437. D.A. Boyd, L. Greengard, M. Brongersma, M.Y. El-Naggar, D.G. Goodwin, Plasmon-assisted chemical vapor deposition. Nano Lett. **6**, 2592–2597 (2006)
438. G.D. Martino, F.B. Michaelis, A.R. Salmon, S. Hofmann, J.J. Baumberg, Controlling nanowire growth by light. Nano Lett. **15**, 7452–7457 (2015)
439. I.L. Violi, J. Gargiulo, C. von Bilderling, E. Cortes, F.D. Stefani, Light-induced polarization-directed growth of optically printed gold nanoparticles. Nano Lett. **16**, 6529–6533 (2016)
440. A. Biswas, N. Lemcoff, O. Shelonchik, D. Yesodi, E. Yehezkel, E.Y. Finestone, A. Upcher, Y. Weizmann, Photothermally heated colloidal synthesis of nanoparticles driven by silica-encapsulated plasmonic heat sources. Nat. Commun. **14** (2023)
441. R. Kamarudheen, G. Kumari, A. Baldi, Plasmon-driven synthesis of individual metal@semiconductor core@shell nanoparticles. Nat. Commun. **11**, 3957 (2020)
442. W. Guo, A.C. Johnston-Peck, Y. Zhang, Y. Hu, J. Huang, W.D. Wei, Cooperation of hot holes and surface adsorbates in plasmon-driven anisotropic growth of gold nanostars. J. Am. Chem. Soc. **142**, 10921–10925 (2020)
443. L.K. Khorashad, L.V. Besteiro, M.A. Correa-Duarte, S. Burger, Z.M. Wang, A.O. Govorov, Hot electrons generated in chiral plasmonic nanocrystals as a mechanism for surface photochemistry and chiral growth. J. Am. Chem. Soc. **142**, 4193–4205 (2020)
444. O. Shelonchik, N. Lemcoff, R. Shimoni, A. Biswas, E. Yehezkel, D. Yesodi, I. Hod, Y. Weizmann, Light-induced MOF synthesis enabling composite photothermal materials. Nat. Commun. **15**, 1154 (2024)
445. L.V. Besteiro, A. Movsesyan, O. Avalos-Ovando, S. Lee, E. Cortes, M.A. Correa-Duarte, Z.M. Wang, A.O. Govorov, Local growth mediated by plasmonic hot carriers: chirality from achiral nanocrystals using circularly polarized light. Nano Lett. **21**, 10315–10324 (2021)
446. M. Sakamoto, M. Hada, W. Ota, F. Uesugi, T. Sato, Localised surface plasmon resonance inducing cooperative Jahn-Teller effect for crystal phase-change in a nanocrystal. Nat. Commun. **14**, 4471 (2023)
447. S.S. Kharintsev, E.I. Battalova, T.A. Mukhametzyanov, A.P. Pushkarev, I.G. Scheblykin, S.V. Makarov, E.O. Potma, D.A. Fishman, Light-controlled multiphase structuring of perovskite crystal enabled by thermoplasmonic metasurface. ACS Nano **17**, 9235–9244 (2023)
448. A.R.C. Baljon, M.O. Robbins, Energy dissipation during rupture of adhesive bonds. Science **271**, 482–484 (1996)
449. F. Patolsky, Y. Weizmann, I. Willner, Actin-based metallic nanowires as bio-nanotransporters. Nat. Mater. **3**, 692–695 (2004)
450. H. Shehla, A. Ishaq, Y. Khan, I. Javed, R. Saira, N. Shahzad, M. Maaza, Ion beam irradiation-induced nano-welding of Ag nanowires. Micro & Nano Letters **11**, 34–37 (2016)
451. J.E. Krzanowski, A transmission electron microscopy study of ultrasonic wire bonding. IEEE Trans. Componen. Hybr. Manuf. Technol. **13**, 176–181 (1989)
452. Z. Lin, X. Gui, Q. Gan, W. Chen, X. Cheng, M. Liu, Y. Zhu, Y. Yang, A. Cao, Z. Tang, In-situ welding carbon nanotubes into a porous solid with super-high compressive strength and fatigue resistance. Sci. Rep. **5**, 11336 (2015)
453. P. Peng, W. Guo, Y. Zhu, L. Liu, G. Zou, Y.N. Zhou, Nanoscale wire bonding of individual Ag nanowires on au substrate at room temperature. Nanomicro Lett **9**, 26 (2017)
454. D.V. Wagle, G.A. Baker, Cold welding: a phenomenon for spontaneous self-healing and shape genesis at the nanoscale. Mater. Horizons **2**, 157–167 (2015)

455. Y. Liu, J. Zhang, H. Gao, Y. Wang, Q. Liu, S. Huang, C.F. Guo, Z. Ren, Capillary-force-induced cold welding in silver-nanowire-based flexible transparent electrodes. Nano Lett. **17**, 1090–1096 (2017)
456. G. Dai, B. Wang, S. Xu, Y. Lu, Y. Shen, Side-to-side cold welding for controllable nanogap formation from "dumbbell" ultrathin gold nanorods. ACS Appl. Mater. Interfaces **8**, 13506–13511 (2016)
457. Y. Lu, J.Y. Huang, C. Wang, S. Sun, J. Lou, Cold welding of ultrathin gold nanowires. Nat. Nanotechnol. **5**, 218–224 (2010)
458. P.E.B. Changsoon Kim, Stephen R. Forres, Micropatterning of organic electronic devices by cold-welding. Science **288**, 831–833 (2000)
459. F. Gao, Z. Gu, Nano-soldering of magnetically aligned three-dimensional nanowire networks. Nanotechnology **21**, 115604 (2010)
460. Z.Y. Gu, H.K. Ye, D.H. Gracias, The bonding of nanowire assemblies using adhesive and solder. Jom-Us **57**, 60–64 (2005)
461. Y. Peng, T. Cullis, B. Inkson, Bottom-up nanoconstruction by the welding of individual metallic nanoobjects using nanoscale solder. Nano Lett. **9**, 91–96 (2009)
462. H. Tohmyoh, T. Imaizumi, H. Hayashi, M. Saka, Welding of Pt nanowires by Joule heating. Scripta Mater. **57**, 953–956 (2007)
463. T.B. Song, Y. Chen, C.H. Chung, Y.M. Yang, B. Bob, H.S. Duan, G. Li, K.N. Tu, Y. Huang, Y. Yang, Nanoscale Joule heating an electromigration enhanced ripening of silver nanowire contacts. ACS Nano **8**, 2804–2811 (2014)
464. R. Chellattoan, V. Lube, G. Lubineau, Toward programmable materials for wearable electronics: electrical welding turns sensors into conductors. Adv. Electron. Mater. **5** (2018)
465. E. Koleva, Electron beam weld parameters and thermal efficiency improvement. Vacuum **77**, 413–421 (2005)
466. M. Terrones, F. Banhart, N. Grobert, J.C. Charlier, H. Terrones, P.M. Ajayan, Molecular junctions by joining single-walled carbon nanotubes. Phys. Rev. Lett. **89**, 075505 (2002)
467. L. Zhang, Y. Tang, Q. Peng, T. Yang, Q. Liu, Y. Wang, Y. Li, C. Du, Y. Sun, L. Cui, F. Yang, T. Shen, Z. Shan, J. Huang, Ceramic nanowelding. Nat. Commun. **9**, 96 (2018)
468. G. González-Rubio, P. Díaz-Núñez, A. Rivera, A. Prada, G. Tardajos, J. González-Izquierdo, L. Bañares, P. Llombart, L.G. Macdowell, M.A. Palafox, L.M. Liz-Marzán, O. Peña-Rodríguez, A. Guerrero-Martínez, Femtosecond laser reshaping yields gold nanorods with ultranarrow surface plasmon resonances. Science **358**, 640–644 (2017)
469. H. Huang, L. Liu, P. Peng, A. Hu, W.W. Duley, Y. Zhou, Controlled joining of Ag nanoparticles with femtosecond laser radiation. J. Appl. Phys. **112** (2012)
470. T. Itoh, Y.S. Yamamoto, H. Tamaru, V. Biju, N. Murase, Y. Ozaki, Excitation laser energy dependence of surface-enhanced fluorescence showing plasmon-induced ultrafast electronic dynamics in dye molecules. Phys. Rev. B **87** (2013)
471. G.V. Hartland, Optical studies of dynamics in noble metal nanostructures. Chem. Rev. **111**, 3858–3887 (2011)
472. J.H. Park, S. Jeong, E.J. Lee, S.S. Lee, J.Y. Seok, M. Yang, Y. Choi, B. Kang, Transversally extended laser plasmonic welding for oxidation-free copper fabrication toward high-fidelity optoelectronics. Chem. Mater. **28**, 4151–4159 (2016)
473. E.C. Garnett, W. Cai, J.J. Cha, F. Mahmood, S.T. Connor, M. Greyson Christoforo, Y. Cui, M.D. McGehee, M.L. Brongersma, Self-limited plasmonic welding of silver nanowire junctions. Nat. Mater. **11**, 241–249 (2012)
474. H. Pan, S.H. Ko, C.P. Grigoropoulos, The coalescence of supported gold nanoparticles induced by nanosecond laser irradiation. Appl. Phys. A **90**, 247–253 (2007)
475. J.H. Lee, B.C. Huynh-Nguyen, E. Ko, J.H. Kim, G.H. Seong, Fabrication of flexible, transparent silver nanowire electrodes for amperometric detection of hydrogen peroxide. Sens. Actuators, B Chem. **224**, 789–797 (2016)
476. R. Rahimi, M. Ochoa, B. Ziaie, Direct laser writing of porous-carbon/silver nanocomposite for flexible electronics. ACS Appl. Mater. Interfaces **8**, 16907–16913 (2016)

477. F. Mafune, J.Y. Kohno, Y. Takeda, T. Kondow, Nanoscale soldering of metal nanoparticles for construction of higher-order structures. J. Am. Chem. Soc. **125**, 1686–1687 (2003)
478. A. Kim, Y. Won, K. Woo, C.H. Kim, J. Moon, Highly transparent low resistance ZnO/Ag nanowire/ZnO composite electrode for thin film solar cells. ACS Nano **7**, 1081–1091 (2013)
479. Q. Nian, M. Saei, Y. Xu, G. Sabyasachi, B. Deng, Y.P. Chen, G.J. Cheng, Crystalline nanojoining silver nanowire percolated networks on flexible substrate. ASC Nano (2015)
480. S.H. Ko, H. Pan, C.P. Grigoropoulos, C.K. Luscombe, J.M.J. Fréchet, D. Poulikakos, Air stable high resolution organic transistors by selective laser sintering of ink-jet printed metal nanoparticles. Appl. Phys. Lett. **90** (2007)
481. S.Y. Ding, J. Yi, J.F. Li, B. Ren, D.Y. Wu, R. Panneerselvam, Z.Q. Tian, Nanostructure-based plasmon-enhanced Raman spectroscopy for surface analysis of materials. Nat. Rev. Mater. **1** (2016)
482. S.Y. Ding, E.M. You, Z.Q. Tian, M. Moskovits, Electromagnetic theories of surface-enhanced Raman spectroscopy. Chem. Soc. Rev. **46**, 4042–4076 (2017)
483. D. Liu, C. Xue, Plasmonic coupling architectures for enhanced photocatalysis. Adv. Mater e2005738 (2021)
484. A. Hardiansyah, A.Y. Chen, H.L. Liao, M.C. Yang, T.Y. Liu, T.Y. Chan, H.M. Tsou, C.Y. Kuo, J.K. Wang, Y.L. Wang, Core-shell of FePt@$SiO_2$-Au magnetic nanoparticles for rapid SERS detection. Nanoscale Res. Lett. **10**, 412 (2015)
485. A. Alkurdi, J. Lombard, F. Detcheverry, S. Merabia, Enhanced heat transfer with metal-dielectric core-shell nanoparticles. Phys. Rev. Appl. **13** (2020)
486. S. Viarbitskaya, A. Cuche, A. Teulle, J. Sharma, C. Girard, A. Arbouet, E. Dujardin, Plasmonic hot printing in gold nanoprisms. ACS Photon. **2**, 744–751 (2015)
487. M. Rahman, K.S. Islam, T.M. Dip, M.F.M. Chowdhury, S.R. Debnath, S.M.M. Hasan, M.S. Sakib, T. Saha, R. Padhye, S. Houshyar, A review on nanomaterial-based additive manufacturing: dynamics in properties, prospects, and challenges. Progr. Addit. Manuf. (2023)
488. C. Zhao, B. Shi, S. Chen, D. Du, T. Sun, B.J. Simonds, K. Fezzaa, A.D. Rollett, Laser melting modes in metal powder bed fusion additive manufacturing. Rev. Modern Phys. **94** (2022)
489. T. Hupfeld, A. Wegner, M. Blanke, C. Doñate-Buendía, V. Sharov, S. Nieskens, M. Piechotta, M. Giese, S. Barcikowski, B. Gökce, Plasmonic seasoning: giving color to desktop laser 3D printed polymers by highly dispersed nanoparticles. Adv. Opt. Mater. **8** (2020)
490. A.W. Powell, A. Stavrinadis, S. Christodoulou, R. Quidant, G. Konstantatos, On-demand activation of photochromic nanoheaters for high color purity 3D printing. Nano Lett. **20**, 3485–3491 (2020)
491. A.W. Powell, A. Stavrinadis, I. de Miguel, G. Konstantatos, R. Quidant, White and brightly colored 3D printing based on resonant photothermal sensitizers. Nano Lett. **18**, 6660–6664 (2018)
492. J. Xue, Z.K. Zhou, Z. Wei, R. Su, J. Lai, J. Li, C. Li, T. Zhang, X.H. Wang, Scalable, full-colour and controllable chromotropic plasmonic printing. Nat. Commun. **6**, 8906 (2015)
493. T. Lee, J. Jang, H. Jeong, J. Rho, Plasmonic- and dielectric-based structural coloring: from fundamentals to practical applications. Nano Converg. **5**, 1 (2018)
494. Y. Zhang, Q. Zhang, X. Ouyang, D.Y. Lei, A.P. Zhang, H.Y. Tam, Ultrafast light-controlled growth of silver nanoparticles for direct plasmonic color printing. ACS Nano **12**, 9913–9921 (2018)
495. X.M. Goh, Y. Zheng, S.J. Tan, L. Zhang, K. Kumar, C.W. Qiu, J.K. Yang, Three-dimensional plasmonic stereoscopic prints in full colour. Nat. Commun. **5**, 5361 (2014)
496. F. Zhang, M. Pu, P. Gao, J. Jin, X. Li, Y. Guo, X. Ma, J. Luo, H. Yu, X. Luo, Simultaneous full-color printing and holography enabled by centimeter-scale plasmonic metasurfaces. Adv. Sci. (Weinh) **7**, 1903156 (2020)
497. S.J. Tan, L. Zhang, D. Zhu, X.M. Goh, Y.M. Wang, K. Kumar, C.W. Qiu, J.K. Yang, Plasmonic color palettes for photorealistic printing with aluminum nanostructures. Nano Lett. **14**, 4023–4029 (2014)

498. X. Zhu, C. Vannahme, E. Hojlund-Nielsen, N.A. Mortensen, A. Kristensen, Plasmonic colour laser printing. Nat. Nanotechnol. **11**, 325–329 (2016)
499. S.N. Chowdhury, P. Nyga, Z.A. Kudyshev, E. Garcia Bravo, A.S. Lagutchev, A.V. Kildishev, V.M. Shalaev, A. Boltasseva, Lithography-free plasmonic color printing with femtosecond laser on semicontinuous silver films. ACS Photon. **8**, 521–530 (2020)
500. Z. Yan, Z. Zhang, W. Wu, X. Ji, S. Sun, Y. Jiang, C.C. Tan, L. Yang, C.T. Chong, C.W. Qiu, R. Zhao, Floating solid-state thin films with dynamic structural colour. Nat. Nanotechnol. **16**, 795–801 (2021)

# Chapter 6
# Summary and Outlook

This study provides a comprehensive review of the critical aspects of thermoplasmonics, including the basic principles, materials, characterizations, and burgeoning applications. Fundamental principles of thermoplasmonics are elucidated, encompassing plasmonic theory, thermodynamics, and thermal-induced physics. The discussion on plasmonic theory encompasses metal electromagnetism, plasmon resonances, and quality factors, providing a solid foundation for understanding the interaction of light with metallic nanostructures. In addition, thermodynamics principles such as heat generation, heating schemes, and heat transfer are outlined, highlighting the key mechanisms responsible for thermal effects in plasmonic systems. Furthermore, the thermal-induced physics section discusses the impact of thermoplasmonics on optical physics, stress wave, mass transport, and chemical reactions, offering insights into the wide implications of thermally induced processes.

Thermoplasmonic materials, a crucial support of this burgeoning field, are reviewed with focus on material selection, design, and fabrication. The selection of materials, including metal nanostructures, metallic derivatives, semiconductor crystals, and 2D materials, is discussed in detail, considering their unique properties and potential applications in thermoplasmonic systems. Material design considerations, including size, shape, and coupling effects, are also explored to provide guidelines for tailoring material properties for specific applications. Moreover, material fabrication methods such as bottom-up colloidal synthesis, top-down lithography, and the fabrication of thin films and complex nanoparticles are discussed, emphasizing the importance of precise and reproducible fabrication techniques for advancing thermoplasmonic research.

Characterization techniques play a crucial role in understanding and optimizing thermoplasmonic systems. Simple analytical models have been introduced to describe the optical response of metal nanoparticles, including retardation effects, without the requirement of large computational resources. Computational methodologies, such as the finite-difference time-domain technique, discrete dipole approximation, boundary element methods, and multiple multipole programs, are highlighted



G. Liu, *Thermoplasmonics*, https://doi.org/10.1007/978-981-97-8332-8_6

for their ability to simulate and analyze the optical and thermal behavior of plasmonic nanostructures. Nanoscale thermometry techniques, such as fluorescent molecular probes, nanoprobes, and Raman spectroscopy, are also discussed, offering insight into the temperature distribution and dynamics at the nanoscale level.

The applications of thermoplasmonics across diverse fields, including life, optoelectronic information science, energy, and chemistry, environmental and material sciences are highlighted. Applications in life science, such as photothermal therapy, drug delivery, and cell biology, demonstrate the potential of thermoplasmonics in advancing medical technologies. Similarly, thermoplasmonics shows promise in energy science applications such as thermophotovoltaics, solar desalination, and anti-icing technologies. Furthermore, its significance in environmental science for water purification, sterilization, and removal of microplastics underscores its potential for addressing pressing environmental challenges. Finally, its application in material science, including plasmonic colloidal synthesis, plasmonic welding, and additive manufacturing, highlights the diverse opportunities for integrating thermoplasmonics into advanced material fabrication processes.

This study addresses concepts that intersect with a multitude of scientific disciplines, reflecting the widespread presence of thermally induced effects across all areas of science. The historical depth and diversity of these interactions have become a catalyst for fostering interdisciplinary synergies. Despite significant progress, several vital facets of thermoplasmonics remain to be elucidated:

- Surface plasmon relaxation heating is a complex process involving the interaction of photons, electrons, and phonons, making the models more intricate. Quantitatively distinguishing between photonic-electric and photonic-thermal contributions remains a key problem to be solved.
- The current description of surface plasmon photothermal effects often relies on classical physics models. By integrating quantum optics, the utilization of thermoplasmonic effects in the realm of quantum optical devices can be further expanded.
- We must examine the impact of the heterogeneous influences of electronic structure and optical response on anisotropic geometries in generating energetic carriers. It is crucial to explore the mechanisms by which energy stored in these carriers disseminates throughout multicomponent systems. Elucidating these core physical principles will enhance our comprehension of energy generation and its propagation across various spatial and temporal domains.
- The characterization and testing of thermal effects in micro/nanostructures mainly focus on macroscopic levels. The actual temperature distribution within the structures during the processes is still unknown, and relevant testing methods have not been fully established. It is thus necessary to develop thermal observation instruments with higher temporal and spatial resolution to visually characterize heat generation, conduction, distribution, and modulation within micro-nanostructures, thereby gaining deeper insights into their physical nature.

- The macroscale thermal-induced phenomena such as mass transport, phase transitions, chemical reactions, and stress waves have been observed in various experiments, yet the precise temperature at the nanostructural level remains ambiguous. Consequently, employing thermal imaging techniques that leverage both near-field and far-field approaches could provide deeper understanding and additional insights.
- In engineering applications, the emergence of single-cell thermal microbiology is poised to expand the horizons of thermoplasmonics and its real-world implementations. In these contexts, plasmonic elements are crucial for their ability to absorb photons and localize heat near surfaces. The accompanying components enhance charge manipulation, essential to controlling plasmon resonance and guiding heat distribution. A thorough grasp of the complex interplays between photons, charge carriers, and phonons requires tailored modeling for each specific application.
- Concerning chemical reactions, surface plasmon resonance effectively transmutes low-energy photons into higher-energy carriers and heat, fostering directed chemical transformations. This phenomenon strives to manage reaction pathways by a careful balance of plasmonic and catalytic attributes. Theoretical analysis of molecular dynamics within excited plasmon states may illustrate strategies to lower-energy barriers and surpass rate-limiting phases in selective catalytic reactions.
- For high-temperature operations, the development of plasmonic components and substrates from refractory materials like molybdenum, nickel, and titanium is critical, given their capacity to endure extreme conditions. The thermal performance may benefit from innovating with materials such as metal composites, polymers, aerogels, and phase change materials. Exploring new plasmonic materials, structures, and fabrication techniques will widen the scope of potential applications.

As the field continues to expand, it is anticipated that further innovations and breakthroughs will emerge, solidifying the pivotal role of thermoplasmonics in shaping the landscape of interdisciplinary research and technology development.

**Declaration of Competing Interest** The author declare that they have no known competing financial interests or personal relationships that could have appeared to influence the work reported in this paper.

**Acknowledgements** This research was supported by the Natural Science Foundation of China (52376143, 52076077).

SPRINGER NATURE

# GPSR Compliance

*The European Union's (EU) General Product Safety Regulation (GPSR) is a set of rules that requires consumer products to be safe and our obligations to ensure this.*

*If you have any concerns about our products, you can contact us on ProductSafety@springernature.com*

In case Publisher is established outside the EU, the EU authorized representative is:

Springer Nature Customer Service Center GmbH
Europaplatz 3
69115 Heidelberg, Germany

The manufacturer's authorised representative in the EU is Springer Nature Customer Service Centre GmbH, Europaplatz 3, 69115 Heidelberg, Germany. If you have any concerns regarding our products, please contact ProductSafety@springernature.com

Printed and bound by CPI Group (UK) Ltd, Croydon, CR0 4YY
01/10/2026
02225858-0001